Molluscan Faunas of the United States East Coast

The species-richness seen in the living molluscan fauna of the East Coast of the United States is due to its composite nature, containing faunal elements from three contiguous molluscan provinces. These include:

- the southern warm-water of the **Carolinian Molluscan Province** (Cape Hatteras, North Carolina to Palm Beach, Florida);
- the central temperate-water of the **Virginian Molluscan Province** (Cape Cod, Massachusetts to Cape Hatteras, North Carolina)
- and the northern cold-water of the **Acadian Molluscan Province** (Labrador, Canada to Cape Cod, Massachusetts).

Each of these provinces has been found to house a large component of endemic species, many illustrated for the first time in this book, *Molluscan Faunas of the United States East Coast: Endemism Along Northeastern North America, Past and Present.* To place the modern East Coast molluscan fauna within a geochronological perspective, we also give detailed descriptions of the predecessor faunas that are contained in the Miocene, Pliocene, and Pleistocene fossil beds of eastern North America. For the first time in any book on American shells, hundreds of extinct species from these Neogene fossil beds are illustrated in separate chapters, describing the marine ecosystems and molluscan faunas that lived during these times. The living and fossil molluscan faunas from the entire coast, altogether, are illustrated with 128 plates, showing some of the most interesting and beautiful shells that ever evolved anywhere in the world's oceans.

For the first time in any book on American malacology, the most prevalent mollusk-bearing ecosystems of the East Coast are described, and their resident molluscan faunas are discussed in detail. Some of these include the immense shoals of the Carolina Bay Scallop (*Argopecten gibbus carolinensis*) and their associated malacofauna, the shallow water *Solenastrea* and *Siderastrea* coral bioherms of Cape Hatteras and offshore areas of the Carolinas, the deep water *Lophelia* coral bioherms, the continental slope mud environments ("Mud Lumps"), coastal lagoons filled with *Spartina* Cordgrass meadows, and rocky intertidal communities in the cold northern waters. This overview of the biodiversity of the coastal faunas can be used as a baseline for any future biological surveys, establishing the composition of the faunas that are now being impacted by global climate changes and higher marine temperatures. The new book is also intended to be a companion to "*Molluscan Faunas of the Gulf of Mexico*: *Endemism in North America's Inland Sea*" (Petuch and Berschauer, 2025).

Over the past 25 million years, the East Coast of the United States has been subject to periodic ecological collapses, triggered by rapid and severe changes in the local marine environments. As shown here for the first time, most of these catastrophes have been caused by extraterrestrial events, such as asteroid impacts and the dimming of the sun by interstellar dust clouds. The climatic degeneration caused by these events decimated the shallow coastal marine ecosystems, especially in the southern part of the region. These extraterrestrial influences are correlated with the abrupt changes seen in the Neogene fossil record and are outlined and summarized on numerous charts, diagrams, and maps that are shown throughout the book.

Molluscan Faunas of the United States East Coast

Endemism Along Northeastern North America, Past and Present

Edward J. Petuch
David P. Berschauer

CRC Press
Taylor & Francis Group
Boca Raton London New York

CRC Press is an imprint of the
Taylor & Francis Group, an **informa** business

Designed cover image: Author's own

First edition published 2027
by CRC Press
2385 NW Executive Center Drive, Suite 320, Boca Raton FL 33431

and by CRC Press
4 Park Square, Milton Park, Abingdon, Oxon, OX14 4RN

CRC Press is an imprint of Taylor & Francis Group, LLC

LCCN 2026015866

ISBN: 978-1-041-22571-3 (hbk)
ISBN: 978-1-041-22573-7 (pbk)
ISBN: 978-1-003-73298-3 (ebk)

DOI: 10.1201/9781003732983

Typeset in Warnock Pro Reguler font
by KnowledgeWorks Global Ltd.

Publisher's note: This book has been prepared from camera-ready copy provided by the authors.

MIX
Paper | Supporting responsible forestry
FSC® C013056

Printed and bound in Great Britain by
TJ Books, Padstow, Cornwall

Dedication

This book is dedicated to the following:

Linda J. Petuch; Eric and Rasa Petuch; Brian Petuch, Kendra Berentsen, and Alice Petuch; and Jennifer Petuch

and

Felicia Weisbrot Berschauer; Morgan, Jeremy, Lincoln, and Brantley Coker; Jonathon, Tawni, Nora, and Emmett Berschauer

and

to the memory of Rachel L. Carson

Table of Contents

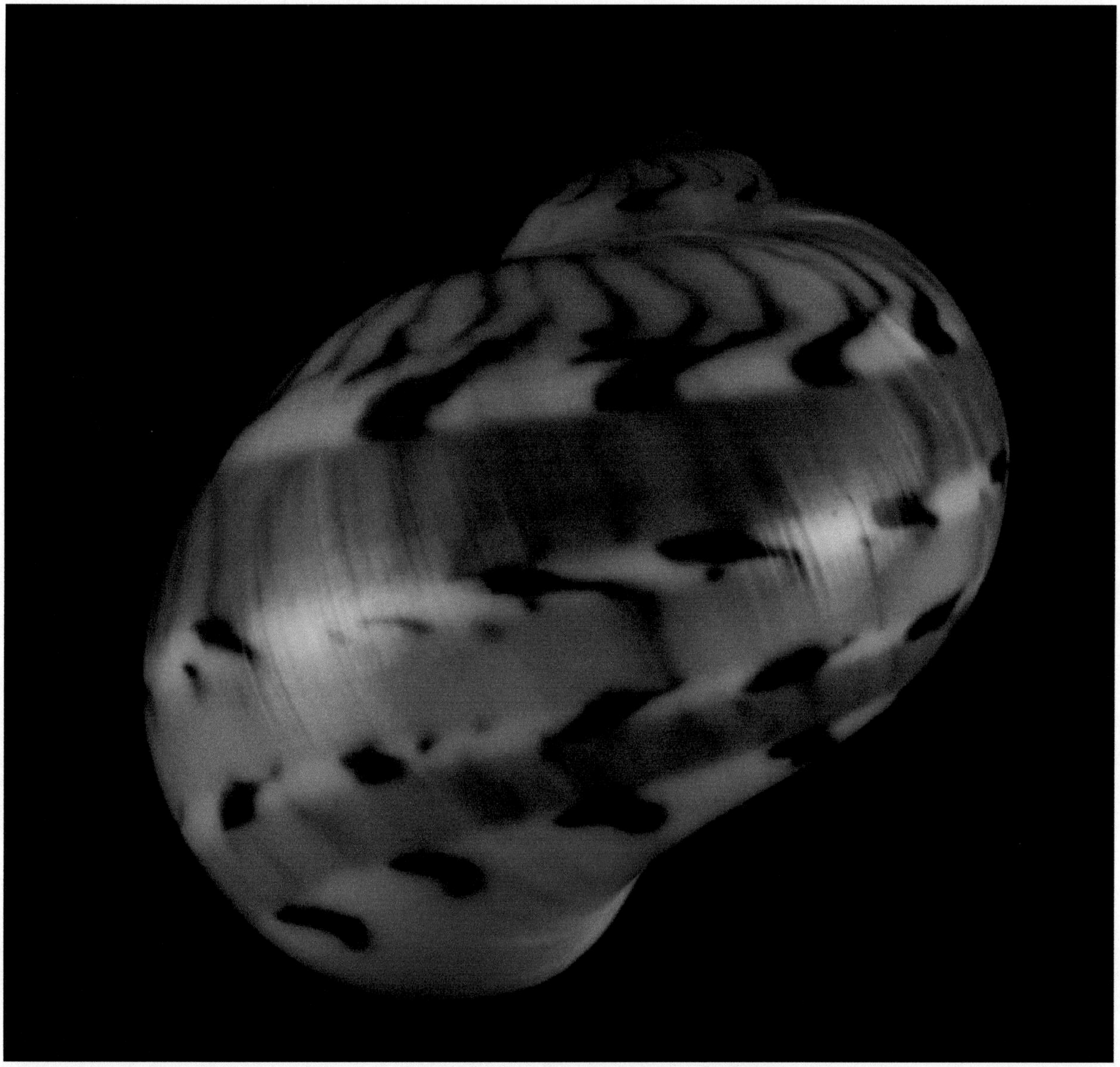

The Colorful Atlantic Natica, *Naticarius canrena* (Linnaeus, 1758), a widespread western Atlantic species.

ACKNOWLEDGMENTS

We thank the following individuals for their help in acquiring important specimens for our research, many of which are illustrated in this book, and for their invaluable help with our field work along the United States East Coast: **Robert Eason**, Paris, Tennessee; **Mark Johnson**, Wilmington, North Carolina; **Eddie Matchett**, Okeechobee, Florida; **Dennis Sargent**, Mount Dora, Florida; **Randy Allamand**, Asheville, North Carolina; **Dr. Kenneth and Wendy Keaton**, Lauderhill, Florida; **Lyle Therriault**, Concord, North Carolina; **Dr. Marvin and Patricia Glickstein**, North Palm Beach, Florida; **Ian Holden**, Halifax, Nova Scotia; **Ida Bucheck**, North Palm Beach, Florida; **Joseph Bucheck, II**, Singer Island, Florida; **Eric Kendrew**, Valrico, Florida; **Emyr Foxhall**, Cardiff, Wales; **Michael Bruggeman**, Atlanta, Georgia; **Gary Leonard**, Atlanta, Georgia; **Harry J. Hyaduck, Sr.**, Atlanta, Georgia; **Donald Asher**, Mechanicsville, Maryland; **Dr. Edward Gerrard**, Hickory, North Carolina; **Linda J. Petuch**, Jupiter, Florida, for assistance with fieldwork in Maryland and North Carolina; **Robert Shaw**, Naturalist, Chancellor Point Natural History area; **Norman Riker**, rural Calvert County, Maryland; **Amy Plummer**, Plum Point, Maryland; **Connie Smith**, St. Leonards, Maryland; **Beth Permelia**, Urbanna, Virginia; **Bonnie Vautrot**, Urbanna, Virginia; **Rachel Germon**, Smithsonian Institution; **Edward Mattingly**, Owings, Maryland; **Joseph Turner**, Baltimore, Maryland; **Vladimir Eismont**, St. Petersburg, Florida; **Dr. J.D. Costlow, Jr.,** Duke University Marine Laboratory, and **the crew of the *R/V Eastward***, Beaufort, North Carolina; **Robert Schmidt**, Solomons, Maryland; **Ralph Chamness**, Aurora, North Carolina; **I.K. ("Tex") Gilmore**, Aurora, North Carolina; **Dahl Daniel**, Urbanna, Virginia; **Bryan Peña**, Portland, Maine; **Marcus and Jose Coltro**, Bergeggi, Italy; **Dr. Geerat Vermeij**, Davis, California; **Douglas N. Shelton**, Mobile, Alabama; **Thomas and Paula Honker**, Delray Beach, Florida; **Frank and Rebecca Hynes**, Aurora, North Carolina; **Tammy Bailey Meyers**, Ormond Beach, Florida; and **Irene White**, Colonial Beach, Virginia; Ernest Roy, Northfield, New Hampshire. Special thanks go to **Charles L. Powell**, San Jose, California and **Roger N. Clark**, Eagle Mountain, Utah, for taking the photos of the holotype of *Bathybela costlowi* shown on Figure 2.38.

Pazinotus stimpsonii **(Dall, 1889), length 8 mm, a widespread western Atlantic muricid that has been collected on the coral bioherms ("coral reefs") off Cape Hatteras, North Carolina.**

INTRODUCTION: East Coast Molluscan Faunas: Past and Present

Unlike the western coast of the contiguous United States, the eastern coast of the United States supports three separate and distinct molluscan faunas. These are contained in separate faunal provinces that are defined by the "50% Rule", where at least half of the resident mollusks are endemic to the geographical limits of the province. These include: the southern warm water **Carolinian Molluscan Province** (Cape Hatteras, North Carolina to Palm Beach, Florida); the central temperate water **Virginian Molluscan Province** (Cape Cod, Massachusetts to Cape Hatteras, North Carolina); and the northern cold water **Acadian Molluscan Province** (Gulf of Maine to Cape Cod, Massachusetts). These three large biogeographical units are described in the following four chapters of this book, along with discussions of their most important ecosystems and illustrations of their characteristic resident mollusks, including the most important endemic index species. For the first time in any book on American malacology, the most prevalent mollusk-bearing ecosystems of the East Coast provinces are described, and their resident molluscan faunas are discussed in detail. Some of the more important of these include the immense shoals of the Carolina Bay Scallop (*Argopecten gibbus carolinensis*) and their associated malacofauna, the shallow water *Solenastrea* and *Siderastrea* coral bioherms of Cape Hatteras and offshore areas of the Carolinas, the deep water *Lophelia* coral bioherms and continental slope mud environments ("Mud Lumps"), coastal lagoons filled with *Spartina* Cordgrass meadows, and rocky intertidal communities in the cold northern waters.

The molluscan faunas that are now living along the United States East Coast are the remnants of much richer and more diverse faunas that existed along that area prior to the Pleistocene Epoch. The impoverished nature of the living faunas of today resulted from the multiple environmental catastrophes that occurred over the past 25 million years. To place the modern East Coast molluscan fauna within a geochronological perspective, we also give detailed descriptions of the predecessor faunas that are contained in the Miocene, Pliocene, and Pleistocene fossil beds of eastern North America. For the first time in any book on American shells, hundreds of extinct species from these Neogene fossil beds are illustrated in the last four chapters, giving insight into the marine ecosystems and molluscan faunas that lived during these times. Over the timespan of the Neogene Period, the East Coast of the United States has been subject to regular catastrophic ecological collapses, triggered by rapid and severe changes in the local marine climates. As shown here for the first time, most of these catastrophes were caused by extraterrestrial events, such as asteroid impacts and the dimming of the sun by interstellar dust clouds. The rapid climatic degeneration caused by these events decimated the shallow coastal marine ecosystems, leading to mass extinctions of the resident molluscan faunas. This was especially apparent in the warmer water southern parts of the region, where entire genera and subfamilies of gastropods and bivalves disappeared abruptly, leaving only impoverished "ghosts" of the original species-rich ecosystems in the subsequent stratigraphic units.

These extraterrestrial influences are correlated here with the abrupt faunal shifts seen in the regional Neogene fossil record and these are outlined and summarized on numerous charts, diagrams, and maps that are shown throughout the book. Some of these catastrophic events include those that were caused by the galactic **Radcliffe Wave** of the middle Langhian Miocene, the **Ries-Steinheim Double Impact** of the early Serravallian Miocene, the **Puffin Impact** of the middle Serravallian Miocene, the **Ewing Impact** of the late Serravallian Miocene, the **Veritas Dust Cloud** of the middle Tortonian Miocene, the **Kara-Kul Impact** of the late Messinian Miocene, and the **Eltanin Impact** of the late Piacenzian Pliocene. Besides causing the extinction of genera and species, these events also destroyed entire molluscan provinces, which had existed for millions of years along the United States East Coast. These ancient biogeographical units and bioregions included the **Transmarian Province** of the Miocene, and the **Caloosahatchian Province** of the Pliocene and early Pleistocene, and the late Pleistocene **Carolinian Province**. A few surviving remnants of these paleoprovinces have managed to survive the catastrophic events and are extant in the present time, lending an archaic element to many of the modern ecosystems.

The patterns of biodiversity of the East Coast molluscan faunas outlined in this book can be used as a baseline for future biological surveys, especially for those marine ecosystems that are now being impacted by global climate changes and higher water temperatures. This new book is also intended to be a companion volume to "*Molluscan Faunas of the Gulf of Mexico*: *Endemism in North America's Inland Sea*" (Petuch and Berschauer, 2025), and together they form a set of volumes that describes the biogeography and marine ecology of the entire Carolinian Molluscan Province. The living and fossil molluscan faunas found from Florida to Maine, altogether, are illustrated here on 128 plates, showing some of the most interesting and beautiful shells that ever evolved anywhere in the world's oceans. In this book, we highlight the endemic species and genera and do not cover the large number of widespread species that occur along with them in the three major biogeographical regions. In this aspect, the book is designed to be an augmentation of previous regional faunal surveys, such as the American Museum of Natural History's "*Guide to Shells: Land Freshwater, and Marine, from Nova Scotia to Florida*" (Emerson and Jacobson, 1976), and "*Seashells of North Carolina, Revised and Expanded Edition*" (Mosher, 2024), adding hundreds more taxa that were missed or were described after these earlier works were published. The emphasis on endemic species is meant to underscore the uniqueness of the United States East Coast malacofauna, setting it aside from all others in the northern and western Atlantic.

Edward J. Petuch, Ph.D. and David P. Berschauer, J.D., 2026

CHAPTER 1. The Carolinian Province

The largest and most species-rich molluscan fauna found along the southeastern United States is contained within the **Carolinian Molluscan Province** (Figure 1.1). Named for the States of North and South Carolina, this immense subtropical biogeographical unit encompasses the entire Gulf of Mexico (see Petuch and Berschauer, 2025) and the eastern coast of the United States as far north as Cape Hatteras, North Carolina. Based upon the percentages of endemic species from around the bioregion, the Carolinian Molluscan Province was found to contain five distinct subsidiary bioregions; four in the Gulf of Mexico and one along the southeastern United States (Petuch, 2013; Petuch and Berschauer, 2022; Petuch and Berschauer, 2025). Using the quantitative methodology and Indices of Provinciality outlined by Petuch (2013), the Carolinian Molluscan Province was found to have a combined level of endemicity that exceeds 75%, far beyond the 50% needed for minimum provincial status, which distinguishes it from the Caribbean, Brazilian, and Paulinian Molluscan Provinces to the south (Petuch and Berschauer, 2022). Besides having numerous endemic species, the Carolinian Province also contains a large array of endemic genera and subgenera, which contribute to the distinctive faunal composition of the local ecosystems.

Within the boundaries of the Carolinian Province, the shorelines are typically dominated by three main types of environments and ecosystems: **1. *Spartina alterniflora* Meadows** (shown here on Figure 1.2), which are characteristic of muddy lagoonal environments in the Georgian, Suwannean, and Texan Subprovinces; **2. Carbonate Sedimentary Environments and Coral Reef Tracts**, which are restricted to the Floridian and Yucatanean Subprovinces (and are not covered in this book; see Petuch and Myers, 2016 and Petuch and Berschauer, 2025 for detailed descriptions of the Florida Coral Reef Tract and its associated mollusks); **3. High-Energy Sand Beaches** (shown here on Figure 1.3), which comprise the majority of near-shore environments in all five Carolinian subprovinces. Composed primarily of quartz sand mixed with carbonate sediments and shell fragments, Carolinian beaches are constantly exposed to heavy surf and high-energy water conditions. Because of this, only a very few species of mollusks can live in this type of habitat. The sand that extends into subtidal depths (Neritic Zone), where there is less turbulence and sediment shifting, offers a more stable substrate and houses a large fauna of characteristic widespread Carolinian species. Some of these sand-dwelling species include the olivid *Americoliva sayana*, the terebrids *Neoterebra dislocata* and *Neoterebra concava*, the strombid *Strombus alatus*, and the conid *Jaspidiconus pfluegeri*. The *Spartina* Cord Grass Meadows, although forming a prominent environment found throughout the northern Gulf of Mexico and along the entire coast between Cape Hatteras and Cape Canaveral, contains an impoverished molluscan fauna, with only a very few resident mollusks. This impoverishment is due primarily to the extreme ecological conditions that typify this intertidal environment; with widely fluctuating temperatures and salinities that change with the seasonal air temperatures and rainfall amounts. Large beds of the Oyster *Crassostrea virginica* occur in the muddy areas between the Cord Grass meadows, offering a hard substrate for predatory Carolinian endemic gastropods such as the large fasciolariid *Cinctura hunteria* and the busyconids *Sinistrofulgur sinistrum* and members of the genus *Fulguropsis* (the Pear Whelks, which comprises 7 distinct species within the Gulf of Mexico and the coastline south of Cape Hatteras).

DOI: 10.1201/9781003732983-1

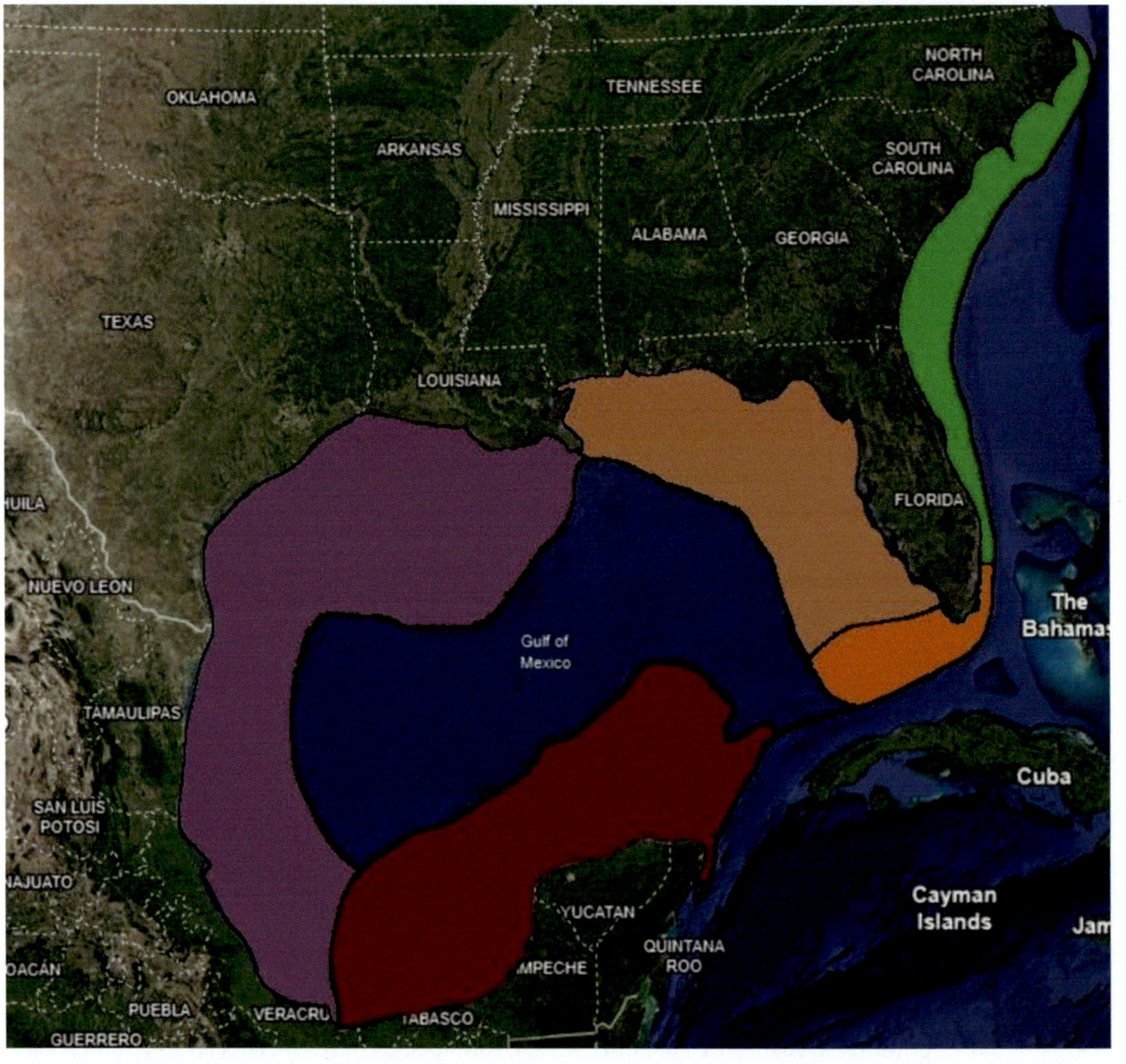

Figure 1.1 Map of the Carolinian Molluscan Province, showing its five Subprovinces.
Green= Georgian Subprovince; **Orange=** Floridian Subprovince; **Tan=** Suwannean Subprovince; **Purple=** Texan Subprovince; **Maroon=** Yucatanean Subprovince. The molluscan faunas of all five subprovinces contain levels of endemism exceeding 25%, as determined by the Indices of Subprovinciality and quantitative methodology outlined in Petuch, 2013.

Within the shallow protected coastal lagoons, which have formed behind the sand barrier islands that line the coast of the eastern United States, large beds of sea grasses grow on the open seafloor between the Cord Grass meadows and oyster bars. Along the eastern coast of the United States, three types of sea grasses make up these underwater meadows: Eel Grass (*Zostera marina*), Shoalgrass (*Halodule wrightii*), and Widgeongrass (*Ruppia maritima*). Farther south along the Florida coast, in the Florida Keys, and throughout the Gulf of Mexico, Turtle Grass (*Thalassia testudinum*) dominates as the sea grass-based environment found in shallow and protected lagoons. Within the Carolinian Province, these grass beds provide a habitat for a characteristic endemic evolutionary radiation of the Bay Scallop genus *Argopecten*, and includes: *Argopecten irradians concentricus* (shown here on Figure 2.7, Chapter 2), which ranges from Cape Hatteras south to Fort Pierce, Florida; *Argopecten irradians taylorae*, which ranges from the Florida Keys (Floridian Subprovince) northward to Mobile Bay, Alabama (Suwannean Subprovince); and

Argopecten irradians amplicostatus, which ranges from the Mississippi River Delta, along Texas, and south to the Bay of Campeche (Texan Subprovince) (these last two Gulf of Mexico subspecies are illustrated, and their morphologies described, in Petuch and Berschauer, 2025). The sea grass beds of the East Coast also house several interesting endemic gastropods such as the modulids *Modulus pacei* and *Trochomodulus calusa foxhalli* (both shown here in Chapter 2), which feed on the algae that coats the individual sea grass blades.

Figure 1.2 View of a Cord Grass (*Spartina alterniflora*) Meadow at Pinckney Island, South Carolina. Salt marshes such as this typically support large aggregations of the gastropods *Littorinopsis irrorata* and *Melampus bidentatus*, which live on the individual emergent grass blades, feeding on encrusting algal films.

The wide muddy sand flats that occur within the coastal lagoons also house an endemic fauna of the venerid bivalve genus *Mercenaria*. The most conspicuous and abundant of these is the very large *Mercenaria campechiensis* (Figure 1.33), which is found throughout the entire Carolinian Province, from Cape Hatteras to the Yucatan Peninsula of Mexico. This widespread Carolinian index species occurs with other more localized species and subspecies, such as *Mercenaria mercenaria notata* in the Carolina lagoon systems and *Mercenaria hartae* in the Palm Beach lagoon systems of southeastern Florida. A similar pattern of sympatry is seen in the Gulf of Mexico, where the widespread *Mercenaria campechiensis* lives together with the smaller *Mercenaria browni* in the Ten Thousand Islands of southwestern Florida and with the smaller *M. texana* in the Texas lagoon systems (see Petuch and Berschauer, 2025 for an overview of this endemic Carolinian Province species complex).

Figure 1.3 View of a typical sandy beach environment at Palmetto Dunes, Hilton Head Island, South Carolina.

Relictual Taxa of the United States East Coast

The eastern coastal waters are now known to have acted as a refugium for several gastropod taxa that were prominent organisms in ancient marine ecosystems. These physiologically plastic mollusks managed to survive through the harsh times of extreme climatic degeneration and mass extinction, and some are now dominant apex predators within modern ecosystems. Primary among these is the buccinoidean family Busyconidae, itself a survivor of the late Cretaceous asteroid impact (Chicxulub Impact) and terminal Mesozoic mass extinction. Having evolved within the Cretaceous Interior Seaway during the Maastrichtian Age, the busyconids underwent a large species radiation during the Miocene, and some of these were ancestral to modern generic groups. Most of the Miocene genera became extinct during the Tortonian-Messinian interval and only a few taxa survived into the early Pliocene (discussed in Chapters 5 and 6 of this book). These Pliocene survivors, in turn, gave rise to a new busyconid species radiation, evolving all the genera and species groups that are now living within the Carolinian Molluscan Province. Relictual survivors such as these are now the principal gastropod predators of the eastern coast of the United States, where they feed primarily on large bivalves such as *Mercenaria*, *Macrocallista*, and *Dinocardium*. These busyconid genera include *Busycon*, *Sinistrofulgur*, *Busycotypus*, and *Fulguropsis* (see Petuch, Myers, and Berschauer, 2015 for a comprehensive review of the families Busyconidae and Echinofulguridae). At present, the largest and most species-rich busyconid fauna is confined to the Gulf of Mexico, where two other genera, *Lindafulgur* and *Busycoarctum*, occur along with the widespread genera *Sinistrofulgur* and *Fulguropsis*.

Another relictual genus that occurs along the East Coast of the United States is the buccinid dwarf whelk genus *Ptychosalpinx*, as typified by the single species, *P. globulus* (Dall, 1889) (Figure 2.38, Chapter 2). This deep-water species, which lives on the outermost edge of the continental shelf and upper Bathyal Zone, in depths of 200-300 m, is the last-living member of a large species radiation that was prominent throughout the Miocene and Pliocene. Several of these ancestral species are shown here in Chapters 5 and 6, demonstrating the rich biodiversity of these odd buccinids during the early and middle Neogene. The modulid genus *Conomodulus*, which lives in offshore red algae beds, is another relictual taxon that is found in deep water along the entire East Coast. The sole living species that occurs there, *Conomodulus lindae* (Petuch, 1987) (Figure 1.7), represents an archaic component within the molluscan assemblages found along the Outer Neritic Zone.

Faunal Submergence

High-latitude, cold-water mollusks can circumvent warm-water ecological barriers in lower latitudes by following the thermocline down into deeper waters. Many species that normally live in shallow water in boreal and arctic areas, in near-freezing water temperatures like those in the Arcadian Molluscan Province, can range much farther south by migrating downslope along the outer edge of the continental shelf and upper Bathyal Zone; a biogeographical pattern referred to here as **Faunal Submergence**. Here, at depths of 200-400 m within the upper Bathyal Zone (continental slope), the water temperatures approximate those found in shallow subtidal depths farther north. For this reason, some of the hardiest boreal species can live offshore of the Carolinas and Georgia, but at very great depths. Some of these Arctic taxa that range into the deep-water areas off the Carolinas include:

Gastropoda
Aporrhaidae
Arrhoges occidentalis mainensis (Johnson, 1926)
Naticidae
Amauropsis islandica (Gmelin, 1791)
Euspira heros (Say, 1822)
Euspira pallida (Broderip and Sowerby, 1829)
Euspira triseriata (Say, 1826)
Colidae
Colus pubescens (Verrill, 1882)
Colus pygmaeus (Gould, 1841)
Colus stimpsoni (Morch, 1867)
Retimohniidae
Retimohnia caelatula (Verrill, 1884)
Retimohnia simplex (Verrill, 1884)
Bivalvia
Pectinidae
Placopecten magellanicus (Gmelin, 1791)
Astartidae
Astarte subequilatera Sowerby II, 1854
Carditidae
Cyclocardia borealis (Conrad, 1831)
Arcticidae
Arctica islandica (Linnaeus, 1767)

The Black Clam, *Arctica islandica*, occurs in dense beds at depths of over 200 m off the Cape Hatteras region, and these form the basis for a large commercial clamming business. The cold-water northern bivalve, *Astarte subequilatera*, also lives in these Black Clam beds and both species, together, are the principal prey items for the large Northern Moon Snail, *Euspira heros*. The Maine Pelican's Foot, *Arrhoges occidentalis* form *mainensis*, also occurs in these clam beds and has been dredged from offshore areas as far south as South Carolina. This ecosystem, which is dominated by *Arctica islandica*, is a classic example of Faunal Submergence along the eastern coastline of the United States. In stark contrast to this deep-water subarctic ecosystem are the Carolina coral reefs, with fully developed Caribbean faunas, which lie directly to the west. Here, tropical warm-water conditions allow Caribbean organisms and Arctic ecosystems to exist within less than 50 km from each other; a situation that is unique to the Carolina coastline.

Caribbean Province Influence in the Carolinian Province

Originating in the Straits of Florida, as a confluence of the **Antilles** and **Florida Currents**, the **Gulf Stream** flows northward along the eastern coast of the United States as far north as Cape Hatteras. From there, the Gulf Stream flows eastward, out into the open North Atlantic, and eventually flows southward along the northwestern coastline of Europe. After hugging the coast of Florida, as far as Cape Canaveral, the Gulf Stream stays far offshore of northern Florida, Georgia, and South Carolina, bathing only the outer edge of the broad continental shelf. This fast-moving tropical oceanic current then comes closest to the Cape Hatteras area, keeping the near-shore shallow banks warm enough to support coral growth all year-round. Along the northern side of Cape Hatteras, colder water conditions exist, the result of the **Shelfbreak Jet**, a southward-flowing countercurrent that originates in the Mid-Atlantic Bight (Cape Cod to Cape

Hatteras) and branches of the arctic **Labrador Current**. The temperate-water marine climate produced by the Shelfbreak Jet prevents warm-water Caribbean organisms from ranging north of Cape Hatteras, which represents the extreme northernmost limit of the Tropical Western Atlantic Region (see Petuch, 2013; Petuch and Berschauer, 2022, for detailed discussions of biogeographical regions).

What are referred to as "coral reefs", from off the Cape Hatteras area, are actually **coral bioherms**, random assemblages of coral species growing together. These bioherms are not true coral reefs because they lack a zonated structure and do not have a deterministic growth pattern (Petuch, 1972; Petuch and Myers, 2016; Petuch and Berschauer, 2025). Although lacking a true reef structure, these coral bioherms offer the preferred substrate for the free-swimming larvae (veligers) of many wide-ranging Caribbean mollusks. Some of these are listed here in the following list, and these include tropical species that also range throughout the Gulf of Mexico, making them major faunal components of the Carolinian Province. Many of these taxa are illustrated and discussed in more detail in Chapter 2 of this book.

CARIBBEAN PROVINCE SPECIES FOUND IN THE CAROLINIAN PROVINCE
Gastropoda
Haliotidae
Haliotis pourtalesii Dall, 1881
Margaritidae
Gaza superba (Dall, 1881)
Turritellidae
Torculoidella acropora (Dall, 1889)
Torcula exoleta (Linnaeus, 1758)
Vermicularia knorrii (Deshayes, 1843)
Melanellidae
Niso hendersoni Bartsch, 1953
Naticidae
Naticarius canrena (Linnaeus, 1758)
Sinum perspectivum (Say, 1831)
Cassidae
Cassis tuberosa (Linnaeus, 1758)
Semicassis cicatricosum (Gmelin, 1791)
Semicassis granulatum (Born, 1778)
Epitoniidae
Alexania floridana (Pilsbry, 1945)
Amaea retifera (Dall, 1889)
Cirsotrema dalli Rehder, 1945
Cirsotrema pilsbryi (McGinty, 1940)
Epitonium krebsii (Morch, 1874)
Stenorhytis pernobilis (Fischer and Bernardi, 1857)
Cypraeidae
Luria cinerea (Gmelin, 1791)
Macrocypraea zebra (Linnaeus, 1758)
Naria acicularis (Gmelin, 1791)
Ovulidae
Cyphoma gibbosum (Linnaeus, 1758)
Cyphoma signatum Pilsbry and McGinty, 1939
Strombidae

Macrostrombus costatus (Gmelin, 1791)

Ranellidae

Linatella caudata poulsenii (Morch, 1877)
Monoplex krebsii (Morch, 1877)
Monoplex parthenopeum (von Salis, 1793)
Monoplex pileare (Linnaeus, 1758)
Ranularia caribbaeum (Clench and Turner, 1957)
Turritriton labiosum (Wood, 1828)

Personiidae

Rhyssema clathrata (Lamarck, 1816)
Rhyssema mcgintyi Emerson and Puffer, 1953

Charoniidae

Charonia variegata (Lamarck, 1816)

Bursidae

Bursa cubaniana (d'Orbigny, 1842)
Lampasopsis thomae (d'Orbigny, 1842)

Tonnidae

Eudolium crosseanum (Monterosato, 1869)
Oocorys abyssorum (Verrill and Smith, 1884)
Oocorys sulcata Fischer, 1883
Tonna galea (Linnaeus, 1758)
Tonna pennata (Morch, 1852

Muricidae

Babelomurex lintoni (Verrill, 1882)
Babelomurex scalariformis (Lamarck, 1822)
Murexiella hidalgoi (Crosse, 1869)
Paziella pazii (Crosse, 1869)
Pazinotus stimpsonii (Dall, 1889)
Phyllonotus pomum (Gmelin, 1791)
Pterynotus phaneus (Dall, 1889)
Siratus beaui (Fischer and Bernardi, 1857)
Stramonita floridana (Conrad, 1837)
Vokesimurex cabritii (Bernardi, 1859)

Fasciolariidae

Fasciolaria tulipa (Linnaeus, 1758)

Pisaniidae

Hesperisternia multangula (Philippi, 1848)

Mitridae

Isara straminea (A. Adams, 1853)
Mitra antillensis Dall, 1889

Harpidae

Oniscidia dennisoni (Reeve, 1842)

Marginellidae

Prunum apicinum (Menke, 1828)
Prunum pruinosum (Hinds, 1844)

Conidae

Conasprelloides villepinii (Fischer and Bernardi, 1857)
Gladioconus mus (Hwass, 1792)
Jaspidiconus mindanus lymani (Clench, 1942)

Terebridae

Neoterebra dislocata (Say, 1822)

Architectonicidae

Architectonica sunderlandi Petuch, 1987 (incorrectly considered a synonym of *A. uruguayana* by Bieler and Petit, 2005)

Turridae

Cryptogemmula periscelida (Dall, 1889)

Bivalvia

Pectinidae

Lindapecten muscosus (Wood, 1828)
Nodipecten fragosus (Conrad, 1849)

As shown in this book, the Carolinian Province is now known to contain a composite molluscan fauna, composed of four distinct temporal and spatial components. These include: **1. Wide-Ranging Carolinian Index Species**, endemic taxa that are found throughout the entire bioregion (some of these are listed next and illustrated in the Iconography at the end of this chapter; with an example being *Triplofusus giganteus*, which is found from Cape Hatteras to Yucatan, Mexico); **2. Species that are Restricted to a Single Subprovince**, endemic index taxa that are only found in one of the five subprovinces of the Carolinian Province (shown here on Figure 1.1; an example is *Siphovasum latiriforme* from the Yucatanean Subprovince); **3. Wide-Ranging Tropical Western Atlantic Species**, species with long-lived planktotrophic larvae and whose ranges extend throughout the Caribbean and Carolinian Provinces (and in a few cases, also the Brazilian Molluscan Province; an example is *Naria acicularis*); **4. Relictual Genera**, that survived previous mass extinctions and are now prominent organisms in Carolinian Province ecosystems (an example is the busyconid genus *Sinistrofulgur*). The following is a listing of the most important species that are found throughout the entire Carolinian Molluscan Province:

WIDE-RANGING CAROLINIAN PROVINCE ENDEMIC SPECIES

Gastropoda

Calliostomatidae

Calliostoma bairdii Verrill and Smith, 1880
Calliostoma euglyptum (A. Adams, 1846)
Calliostoma marionae Dall, 1906
Calliostoma sayanum Dall, 1889
Calliostoma yucatecanum Dall, 1881

Solariellidae

Solariella lacunella (Dall, 1881)

Cyclostrematidae

Cyclostrema tortugana (Dall, 1927)

Neritidae

Nerita (Theliostyla) lindae Petuch, 1988

Modulidae

Conomodulus lindae Petuch, 1987
Modulus kaicherae Petuch, 1987

Turritellidae

Turritellopsis floridana Dall, 1927

Cypraeidae

Macrocypraea (Lorenzicypraea) cervus (Linnaeus, 1771)

Ovulidae

Cyphoma mcgintyi Pilsbry, 1939
Simnialena uniplicata (Sowerby II, 1849)

Cassidae
Cassis spinella Clench, 1944
Personiidae
Distorsio (Rhyssema) clathrata (Lamarck, 1816)
Ficidae
Ficus papyratia (Say, 1822)
Muricidae
Chicoreus dilectus (A. Adams, 1855)
Eupleura sulcidentata Dall, 1890
Favartia cellulosa (Conrad, 1846)
Favartia levicula (Dall, 1889)
Hexaplex fulvescens (Sowerby I, 1834)
Pteropurpura bequaerti (Clench and Farfante, 1945)
Vokesimurex morrisoni Petuch and Sargent, 2011
Vokesimurex rubidus (F.C. Baker, 1897)
Fasciolariidae
Aristofusus excavatus (Sowerby II, 1880) (= *eucosmius*)
Cinctura hunteria (Perry, 1811)
Triplofusus giganteus (Kiener, 1840)
Triplofusus giganteus form *reevei* (Jonas, 1850)
Nassariidae
Antillophos virginiae (Schwengel, 1942)
Pisaniidae
Engina corinnae Crovo, 1971
Volutidae
Clenchina gouldiana (Dall, 1887)
Conidae
Conasprelloides stimpsoni (Dall, 1902)
Dalliconus mcgintyi (Pilsbry, 1955)
Dauciconus amphiurgus (Dall, 1889)
Jaspidiconus pfluegeri Petuch, 2004
Kohniconus delessertii (Recluz, 1843)
Lindaconus atlanticus (Clench, 1942)
Terebridae
Neoterebra onslowensis (Petuch, 1974)
Bivalvia
Pectinidae
Euvola marshallae Petuch and Myers, 2014 (found only as far north as NE Florida along the east coast)

These wide-ranging Carolinian Province species are illustrated in the following iconography.

Poulsen's Triton, *Linatella caudata poulseni* (Morch, 1877), the western Atlantic subspecies of the Indo-Pacific *Linatella caudata caudata* (Gmelin, 1791). Within the Georgian Subprovince, this rarely seen subspecies lives offshore on the beds of the scallop *Argopecten gibbus carolinensis*.

Iconography of Widespread Carolinian Index Species

Figure 1.4 Widespread Carolinian and Caribbean Calliostomatids.
A= *Calliostoma bairdii* Verrill and Smith, 1889, height 18 mm; **B**= *Calliostoma sayanum* Dall, 1889, holotype, diameter 39 mm; **C**= *Calliostoma marionae* Dall, 1906, height 21 mm; **D**= *Calliostoma euglyptum* (A. Adams, 1854), height 25 mm; **E**= *Calliostoma roseolum* Dall, 1880, height 11 mm; **F**= *Calliostoma jujubinum* (Gmelin, 1791), height 14 mm. These widespread Carolinian *Calliostoma* species are found on offshore coral bioherms and rocky seafloors from North Carolina to Florida, with some species ranging into the Caribbean Sea.

Figure 1.5 Widespread Carolinian Deer Cowrie.

A, B= *Macrocypraea (Lorenzicypraea) cervus* (Linnaeus, 1771), length 101 mm. Found throughout the entire Carolinian Province, from Cape Hatteras, North Carolina south to the Florida Keys and throughout the entire Gulf of Mexico. The Deer Cowrie is especially common on the coral bioherms off Cape Hatteras and Cape Lookout. It is a generalist feeder, eating sponges, hydroids, algae, and even carrion. An isolated dwarf subspecies, *Macrocypraea cervus lindseyi* Petuch, 2013, was discovered to be living along southeastern Cuba.

Figure 1.6 Widespread Carolinian Ovulids.
A, B= *Cyphoma mcgintyi* Pilsbry, 1939, length 28 mm, found on gorgonians on offshore hard bottom areas throughout the Carolinian Province; **C, D=** *Cyphoma gibbosum* (Linnaeus, 1758), length 29 mm, a Caribbean and Florida Keys species that has been discovered on the coral bioherms off North Carolina; **E, F=** *Cyphoma gibbulum* (Cate, 1978), length 13.5 mm, a rare deep-water species found along the east coast of Florida and may range farther northward; **G, H=** *Simnialena uniplicata* (Sowerby II, 1849), length 15 mm, found on *Leptogorgia* gorgonians throughout the entire Carolinian Province and was recently found to range to the coast of Virginia, making it the only ovulid found in the Virginian Province.

Figure 1.7 Linda's Modulus Shell.
A, B= *Conomodulus lindae* (Petuch, 1987), height 9 mm; **C, D**= *Conomodulus lindae* (Petuch, 1987), height 8 mm; **E, F**= *Conomodulus lindae* (Petuch, 1987), height 11 mm. Found in depths of 40-50 m all along the east coast, from offshore South Carolina to Palm Beach, Florida and the Florida Keys. Off Mayport, northeastern Florida, *C. lindae* lives on red algae beds, growing on exposed rock platforms, in 20 m depth.

Figure 1.8 Small-Spined Helmet Shell.
A, B= *Cassis spinella* Clench, 1944, length 263 mm. Endemic to the Carolinian Province and not found in the Caribbean Sea region to the south. This large species lives together with the Caribbean Province *Cassis madagascariensis* Lamarck, 1822 in the Florida Keys and along the Yucatan Peninsula of Mexico. A dwarf form of *Cassis spinella* occurs in Onslow Bay, North Carolina and is frequently collected by commercial scallop dredging operations.

Figure 1.9 Common Fig Shell, *Ficus papyratia.*
A, B= *Ficus papyratia* (Say, 1822), length 86 mm; **C, D=** *Ficus papyratia* (Say, 1822), length 75 mm. This common sand-dwelling species is found throughout the entire Carolinian Province, from Cape Hatteras, North Carolina south to the Florida Keys and throughout the entire Gulf of Mexico. It has been misidentified, in several recent publications (such as Mosher, 2024), as "*Ficus ficus* (Linnaeus, 1758)", which is an Indian Ocean and Western Pacific species and is not found in the Atlantic Ocean. *Ficus communis* (Roding, 1798) is also a synonym of the Indo-Pacific *Ficus ficus.*

Figure 1.10 Great White Murex from offshore hard-bottom areas.
A, B= *Hexaplex fulvescens* (Sowerby I, 1834), length 144 mm. Found throughout the Carolinian Province, from North Carolina to the Florida Keys and throughout the Gulf of Mexico. This is the largest muricid found along the eastern United States and is an index species for the Carolinian Molluscan Province.

Figure 1.11 Widespread Carolinian Muricid Gastropods.
A, B= *Favartia levicula* (Dall, 1889), length 17 mm; **C, D**= *Eupleura sulcidentata* Dall, 1890, length 22 mm; **E, F**= *Vokesimurex rubidus* (F.C. Baker, 1897), length 35 mm; **G, H**= *Favartia cellulosa* (Conrad, 1846), length 15 mm. These four muricid species are classic Carolinian Province taxa and are found throughout the Gulf of Mexico and along the United States East Coast as far as Cape Hatteras. They are common on the coral bioherms and hard seafloor areas off North Carolina.

Figure 1.12 Carolinian Lace Murex from sea grass beds.
A, B= *Chicoreus dilectus* (A. Adams, 1855), length 51 mm; **C, D**= *Chicoreus dilectus* (A. Adams, 1855), dark-color form, length 56 mm. This ornate muricid species is a classic Carolinian Province index taxon and is found throughout the Gulf of Mexico and along the United States East Coast as far as Cape Hatteras. It is common on the coral bioherms and hard seafloor areas off North Carolina. In her new book on North Carolina shells, Mosher, 2024, erroneously refers to this classic Carolinian species as "*Chicoreus florifer* (Reeve, 1846)"; but that name is now known to refer to a different species that is confined to the Caribbean Province and that is not found in the Carolinian Province. *Chicoreus florifer* is a much broader, wider shell with proportionally larger frondose varices and a lower spire, quite different from the slender and high-spired *C. dilectus.*

Figure 1.13 Bequaert's Winged Murex from deep-water areas.
A, B= *Pteropurpura bequaerti* (Clench and Farfante, 1953), length 36 mm; **C, D=** *Pteropurpura bequaerti* (Clench and Farfante, 1953), length 34 mm. This unusual winged muricid species is a classic Carolinian Province index taxon and is found throughout the Gulf of Mexico and along the United States East Coast as far as Cape Hatteras. Along the eastern United States, Bequaert's Winged Murex lives on rocky outcrops and *Lophelia* coral bioherms, in depths of around 200 m, along the outer edge of the continental shelf.

Figure 1.14 The Horse Conch, a classic Carolinian Province index species.
A, B= *Triplofusus giganteus* (Kiener, 1840), length 271 mm. This large fasciolariid species is a classic Carolinian Province index taxon and is found throughout the Gulf of Mexico and along the United States East Coast as far as Cape Hatteras. It is the largest gastropod of the Carolinian Province, with some especially gigantic specimens are almost 1 m in length. *Triplofusus papillosus* (Sowerby I, 1825), named from a juvenile specimen, may be the correct name for this species, but the taxonomy is still in debate.

Figure 1.15 Reeve's Horse Conch.
A, B= *Triplofusus giganteus* form *reevei* (Jonas, 1850), length 278 mm. This knobless variety of the Horse Conch appears in every population of the Horse Conch found throughout its entire range.

Figure 1.16 Common Banded Tulip Shell.
A, B= *Cinctura hunteria* (Perry, 1811), length 97 mm; **C, D=** *Cinctura hunteria* (Perry, 1811), length 63.4 mm; **E, F=** *Cinctura hunteria* (Perry, 1811), length 63.2 mm. A very widespread Carolinian Province species, ranging from Cape Hatteras, North Carolina to western Florida. The specimen shown here as A and B was collected at low tide along Shackleford Island, off Beaufort, Carteret County, North Carolina, at the northernmost end of the species' range.

Figure 1.17 The Left-Handed Whelk.
A, B= *Sinistrofulgur sinistrum* (Hollister, 1958), length 176 mm. This specimen is typical of *sinistrum* specimens found along the eastern coast of the United States, where the species ranges from North Carolina south to the Florida Keys and all along western Florida and the northern Gulf of Mexico. The East Coast specimens are more inflated and pyriform than those found in the Gulf of Mexico and may represent an undescribed subspecies.

Figure 1.18 Gould's Volute.
A, B= *Clenchina gouldiana* (Dall, 1887), holotype, length 62.3 mm, off Cape Fear, North Carolina; **C, D=** *Clenchina gouldiana* (Dall, 1887), brown-banded color form, length 54 mm; **E, F=** *Clenchina gouldiana* (Dall, 1887), form with faint banding, length 61 mm. A widespread Carolinian Province species that ranges throughout the Gulf of Mexico and northward to Cape Hatteras, North Carolina. This is the most wide-ranging volute in the Carolinian Province. The name "*Scaphella bullisi*" has been applied to unbanded specimens like the holotype, but that taxon was never published and is now considered a *nomen nudum* and is not valid.

Figure 1.19 Georgia Volute.
A, B= *Rehderia georgiana* (Clench, 1946), length 80 mm; **C, D=** *Rehderia georgiana* (Clench, 1946), juvenile specimen showing the strong longitudinal ribs on the early whorls of the spire, length 30 mm. A widespread volute found all along the East Coast, from Cape Hatteras, North Carolina south to Key Largo, Monroe County, Florida.

Figure 1.20 Widespread Carolinian Province Cassid Gastropods.
A, B= *Semicassis granulata* (Born, 1778), high-spired variant, length 72 mm; **C, D=** *Semicassis granulata* (Born, 1778), dark-banded color form, length 76 mm; **E, F=** *Semicassis cicatricosa* (Gmelin, 1791), length 32 mm. These two small cassids range from North Carolina south into the Caribbean Sea and northern Brazil. *Semicassis granulata*, the Scotch Bonnet, is the State Shell of North Carolina.

Figure 1.21 Widespread Carolinian Province Gastropods.
A, B= *Vokesimurex morrisoni* Petuch and Sargent, 2011, length 48 mm (often incorrectly referred to as "*Vokesimurex bellegladeensis*", but that species is now known to be an early Pleistocene fossil from the Bermont Formation of southern Florida); **C, D**= *Cancellaria reticulata* (Linnaeus, 1767), length 32 mm; **E, F**= *Hesperisternia multangulus* (Philippi, 1848), length 26 mm; **G, H**= *Strombus alatus* (Gmelin, 1791), length 96 mm; **I**= *Strombus alatus* (Gmelin, 1791), length 93 mm. These gastropods are classic Carolinian Province taxa and are found throughout the Gulf of Mexico and along the United States East Coast as far as Cape Hatteras. *Vokesimurex morrisoni* is commonly collected in the offshore *Argopecten gibbus carolinensis* beds.

Figure 1.22 Delessert's Cone Shell.
A, B= *Kohniconus delessertii* (Recluz, 1843), length 82 mm; **C, D=** *Kohniconus delessertii* (Recluz, 1843), length 55 mm; **E, F=** *Kohniconus delessertii* (Recluz, 1843), dark-color form, length 39 mm. A widespread Carolinian Province species that ranges from the Gulf of Mexico northward to Cape Hatteras, North Carolina. It is especially common in the scallop beds in Onslow Bay, off Carteret and Onslow Counties, North Carolina.

Figure 1.23 Julia's Cone Shell.
A, B= *Dauciconus amphiurgus* (Dall, 1889), length 31 mm, colorful variant named *juliae* Clench, 1942; **C, D=** *Dauciconus amphiurgus* (Dall, 1889), typical form, length 32 mm; **E, F=** *Dauciconus amphiurgus* (Dall, 1889), length 26 mm. All three specimens were dredged by scallop boats, north of Cape Canaveral, Brevard County, Florida and the species ranges throughout the entire Carolinian Province, from North Carolina to the Gulf of Mexico. Large and very colorful specimens have been dredged by the scallop boats in Onslow Bay, North Carolina.

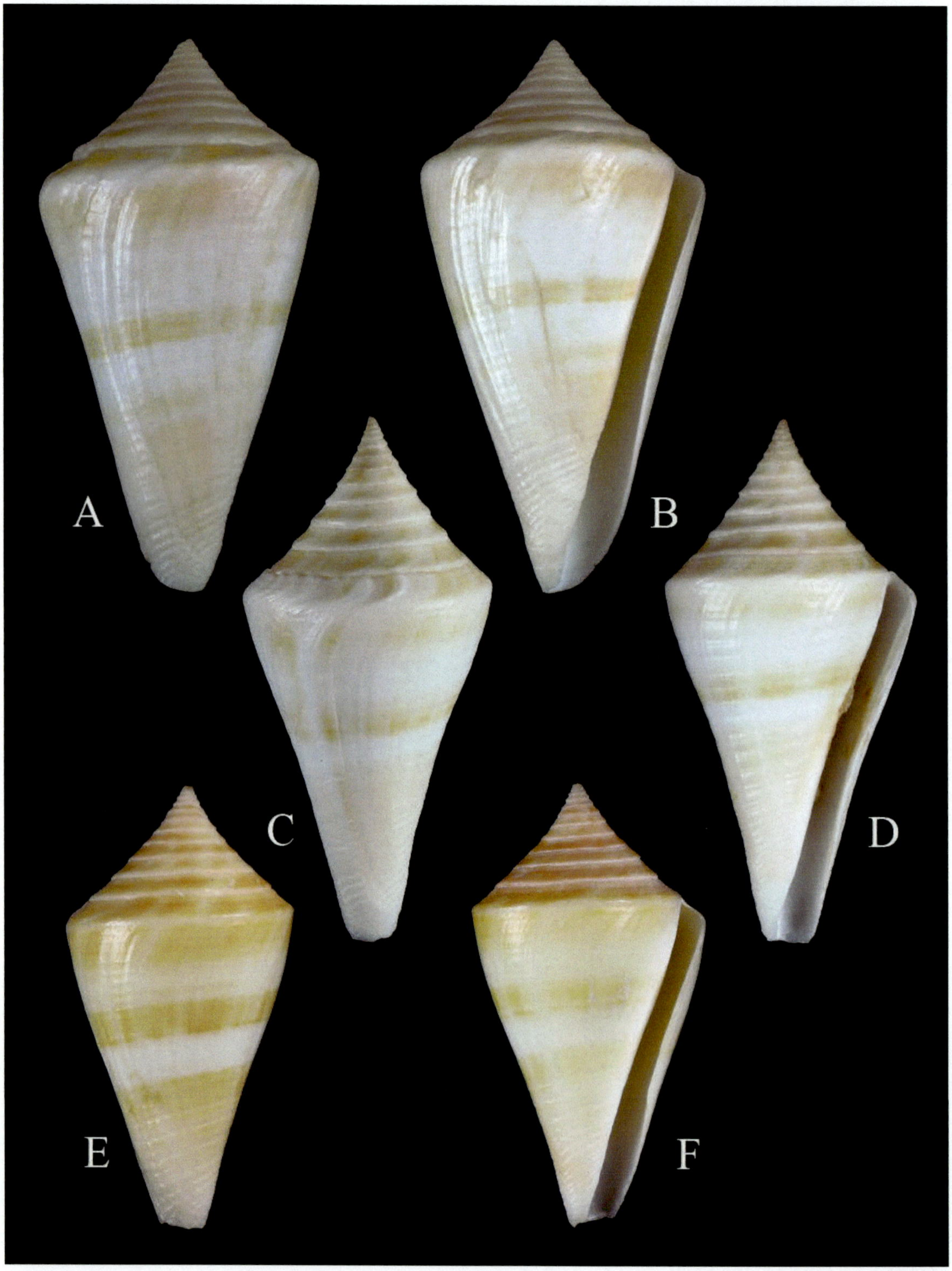

Figure 1.24 Stimpson's Cone Shell.
A, B= *Conasprelloides stimpsoni* (1902), length 49 mm; **C, D=** *Conasprelloides stimpsoni* (1902), heavily ribbed variant that is similar to Dall's holotype, length 45 mm; **E, F=** *Conasprelloides stimpsoni* (1902), darkly banded color form, length 38 mm. A widespread Carolinian Province species that ranges from the Gulf of Mexico northward to Cape Hatteras, North Carolina. It is especially common in the scallop beds in Onslow Bay, off Carteret and Onslow Counties, North Carolina.

Figure 1.25 McGinty's Cone Shell.

A, B= *Dalliconus mcgintyi* (Pilsbry, 1955), length 68 mm; **C, D=** *Dalliconus mcgintyi* (Pilsbry, 1955), length 60 mm; **E=** *Dalliconus mcgintyi* (Pilsbry, 1955), holotype, length 70 mm. This elongated cone shell is a classic Carolinian Province index taxon and is found throughout the Gulf of Mexico and along the United States East Coast as far as Cape Hatteras. Along the eastern United States, McGinty's Cone Shell lives in depths of around 200-300 m along the outer edge of the continental shelf and upper continental slope (Bathyal Zone), often in association with the deep water *Lophelia* coral bioherms. Beautiful and large specimens have been dredged off the Charleston Mud Lumps.

Figure 1.26 Pflueger's Cone Shell.
A, B= *Jaspidiconus pfluegeri* Petuch, 2004, length 25 mm, holotype, **C, D=** *Jaspidiconus pfluegeri* Petuch, 2004, length 23 mm; **E, F=** *Jaspidiconus pfluegeri* Petuch, 2004, length 22 mm. All three specimens were collected in Lake Worth Lagoon, Palm Beach County, Florida. The species also ranges up the East Coast as far as North Carolina, where specimens were collected on the coral bioherms off Cape Lookout. Although morphologically distinct, this conid was erroneously referred to as "*Conus stearnsi*" in many previous works on East Coast shells (such as Mosher, 2024:189); but that species is now known to be endemic to western Florida (see Petuch and Berschauer, 2025, where an entire plate of *stearnsi* is presented, showing several color forms).

Figure 1.27 Cone Shells of the East Coast coral reefs.
A, B= *Jaspidiconus mindanus* form *lymani* (Clench, 1942), length 34 mm; **C, D=** *Gladioconus mus* (Hwass, 1792), length 29 mm; **E, F=** *Atlanticonus granulatus* (Linnaeus, 1758), length 65 mm; **G, H=** *Chelyconus testudinarius* (Hwass, 1792), length 68 mm. These wide-ranging Caribbean and western Atlantic species have been collected on coral bioherms all along the East Coast, from North Carolina south to Florida. Along the Atlantic Coast of the United States, the Glory-of-the-Atlantic Cone (*Atlanticonus granulatus*) has only been found on offshore coral reefs from Cape Canaveral to the Florida Keys. The *lymani* color form of *Jaspidiconus mindanus* has been collected on the coral bioherms off North Carolina. In the new book on North Carolina shells (Mosher, 2024:188), *Chelyconus testudinarius* was identified as "*Conus ermineus* Born, 1778", but that name actually refers to a West African-Eastern Atlantic endemic species. The correct name for the western Atlantic species is *Chelyconus testudinarius* (Hwass, 1792).

Figure 1.28 Widespread Carolinian Province Auger Shells.
A= *Neoterebra concava* (Say, 1827), length 19 mm; **B, C=** *Neoterebra dislocata* (Say, 1822), length 32 mm; **D, E=** *Cinguloterebra floridana* (Dall, 1889), length 54 mm; **F, G=** *Neoterebra protexta* (Conrad, 1845), length 24 mm. These four Auger Shells are found throughout the entire Carolinian Province, from North Carolina south to the Florida Keys and throughout the entire Gulf of Mexico.

Figure 1.29 Onslow Bay Auger Shell.
A, B= *Neoterebra onslowensis* (Petuch, 1974), length 31 mm; **C, D**= *Neoterebra onslowensis* (Petuch, 1974), length 32 mm; **E, F**= *Neoterebra onslowensis* (Petuch, 1974), length 24 mm; **G**= *Neoterebra onslowensis* (Petuch, 1974), paratype, length 40 mm. Originally described from specimens collected by scallop fishermen working in Onslow Bay, North Carolina (Petuch, 1974a). Since then, this distinctive Auger Shell, with its fine-ribbed shell sculpture, has been collected in the Florida Keys and along western Florida, demonstrating that it has a much larger range than originally thought. Specimens from North Carolina are larger and darker than those from the Florida Keys and western Florida.

Figure 1.30 Widespread Scallops of the Carolinian Province.
A, B= *Lindapecten muscosus* (Wood, 1828), length 37 mm; **C, D=** *Aequipecten glyptus* (Verrill, 1882), length 43 mm; **E=** *Cryptopecten phrygium* (Dall, 1886), length 45 mm; **F, G=** *Argopecten nucleus* (Born, 1778), length 35 mm; **H=** *Spathochlamys benedicti* (Verrill and Bush, 1897), length 16 mm. These wide-ranging scallops live offshore in deeper water throughout the Carolinian Province and are rarely collected as beach specimens.

Figure 1.31 Large Scallops from offshore sand seafloors.
A, B= *Nodipecten fragosus* (Conrad, 1849), length 79 mm; **C, D=** *Nodipecten fragosus* (Conrad, 1849), length 80 mm; **E, F=** *Euvola raveneli* (Dall, 1898), length 51 mm. These two scallops are restricted to the Carolinian Province and are found throughout the Gulf of Mexico and northward as far as Cape Hatteras, North Carolina.

Figure 1.32 Marshall's Paper Scallop.
A, B= *Euvola marshallae* Petuch and Myers, 2014, top and bottom valves, length 46 mm; **C, D**= *Euvola marshallae* Petuch and Myers, 2014, top and bottom valves, length 52 mm. This paper-thin deep-water scallop is most abundant and widespread within the Gulf of Mexico (see Petuch and Berschauer, 2025: 44), but a few specimens have been collected from deep water areas (300 m depths) off northeastern Florida. This interesting Carolinian Province endemic has been erroneously referred to as "*Amusium papyraceum* Gabb, 1873" by many previous workers and government fisheries agencies. That name, however, is now known to be an early Miocene (Burdigalian Age) fossil species that was collected in the Baitoa Formation of the Dominican Republic.

Figure 1.33 Campeche Venus Clam.
A, B= *Mercenaria campechiensis* (Gmelin, 1791), length 132 mm; **C=** *Mercenaria campechiensis* (Gmelin, 1791), length 102 mm. This large venerid bivalve is restricted to the Carolinian Province and is found throughout the Gulf of Mexico and northward to off Cape Hatteras, North Carolina.

Figure 1.34. Angel Wing Clam. *Cyrtopleura costata* (Linnaeus, 1758), a deep burrowing bivalve found on intertidal mud flats from the Gulf of Mexico north to Cape Cod, Massachusetts. Their fragile shells, and habit of burrowing down to depths of over 1 m, make them very hard to collect.

CHAPTER 2. The Georgian Subprovince

Along the eastern coast of the United States, the richest molluscan fauna is contained within the Georgian Subprovince of the Carolinian Province (Figure 2.1). Named for the State of Georgia, this large bioregion extends from Cape Hatteras, North Carolina south to Palm Beach, Florida and houses some of the most unusual and distinctive molluscan assemblages found anywhere within the Carolinian Province. This single subprovince, alone, is larger than the combined Virginian and Acadian Provinces to the north and encompasses roughly half the marine environments found along the eastern United States. Of the five subprovinces of the Carolinian Province, the Georgian has the lowest biodiversity, containing only a fraction of the species-richness seen in the Gulf of Mexico subprovinces. North of Cape Canaveral, this impoverishment is the result of low water temperatures during the winter months, preventing tropical organisms from becoming established.

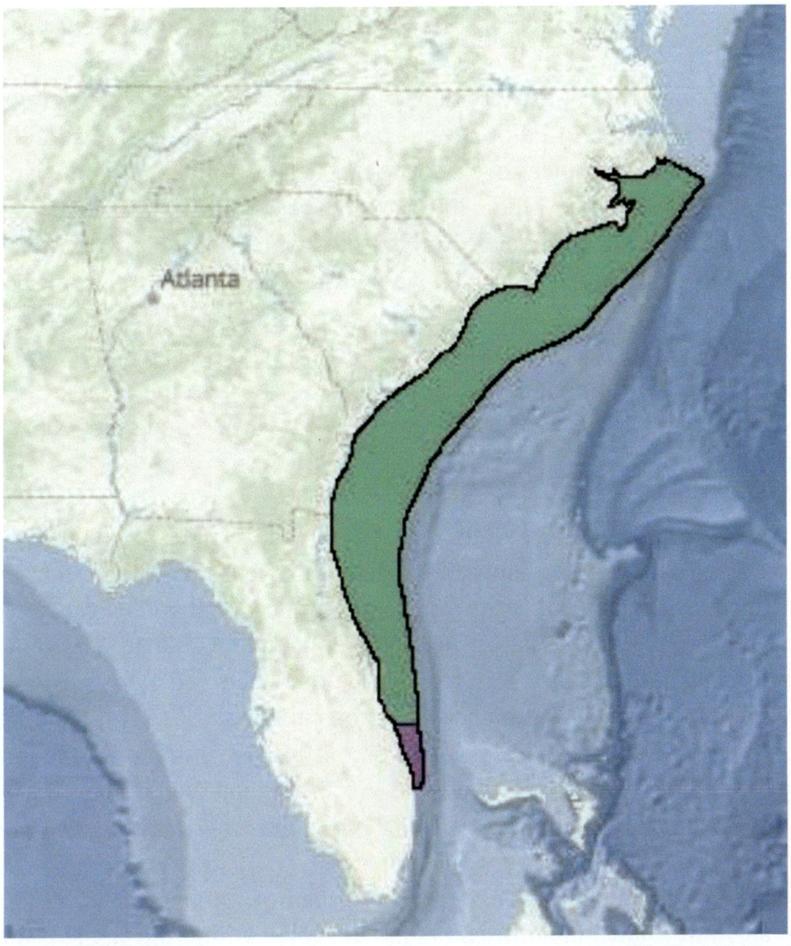

Figure 2.1 Map of the Georgian Subprovince, showing the Palm Beach Infraprovince (purple).

Endemism in the Coastal Lagoon Systems

DOI: 10.1201/9781003732983-2

Much of the coastline of the eastern United States is composed of a series of thin, elongated sand barrier islands and their adjacent sheltered lagoons. The sand barrier islands protect the lagoonal areas from any heavy wave action and provide a stable, low-energy environment for large sea grass beds, salt marsh Cord Grass meadows, oyster banks, and organic-rich mud flats. Typically, the salinity of these areas varies seasonally, with lowered salinity during the summer, and periods of heavy rain, and high salinity during the winter, with decreased rainfall. Likewise, the water temperature within the coastal lagoons alternates between hot, during the summer months, and cold, sometimes near freezing, in the winter. These constantly fluctuating environmental conditions are too harsh for most shallow-water marine animals and only a few physiologically plastic species can live in areas like this and thrive. For this reason, the coastal lagoons all along the Eastern Seaboard contain a very impoverished molluscan fauna, often composed of only a handful of species.

Because of the latitudinal temperature gradient, the coastal lagoons of the Georgian Subprovince break into two main ecoregions: a southern component that is dominated by warm, subtropical environments (Palm Beach, Florida north to Cape Canaveral, Florida) and a northern component that is dominated by temperate environments (Cape Canaveral, Florida to Cape Hatteras, North Carolina). The coastal water temperatures of both ecoregions are completely dependent on the proximity to the tropical Gulf Stream current. In the south (Cape Canaveral to Palm Beach), the Gulf Stream hugs the shoreline and keeps the coastal environments warm and subtropical even during the coldest winter months. North of the Jupiter Inlet, in northern Palm Beach County, the Gulf Stream veers farther offshore, allowing a colder countercurrent to dominate the marine climate during the winter months. This mixed oceanography extends northward to near Fort Pierce, Florida, where the last Caribbean tropical faunal elements disappear from the nearshore areas. The warm Gulf Stream remains far offshore on the wide continental shelf off northern Florida, Georgia, and South Carolina, allowing the coastal waters to fall below tropical conditions in the winter. The Gulf Stream again comes close to shore in the areas off Cape Lookout and Cape Hatteras in North Carolina, where the water remains warm enough to support coral growth (discussed later in this chapter).

The Palm Beach lagoon systems are unique along the eastern United States in that the Gulf Stream water flows inland into the coastal lagoons, producing landlocked tropical marine environments. As the fast-moving Gulf Stream flows over shallow areas close to shore, frictional forces cause numerous spin-off eddies to form along the main current path. These spinning, donut-shaped water masses often flow through the entrance to the Palm Beach Inlet, between Palm Beach and Singer Islands, and bathe the lagoonal islands, like Peanut Island, in permanent high tropical conditions. The planktotrophic larvae of many West Indian mollusks, including the giant strombid *Aliger gigas* and the ranellid *Ranularia caribbaea*, ride these Gulf Stream eddies into the Lake Worth Lagoon, where they settle out and become established as important components within the resident Turtle Grass (*Thalassia testudinum*) ecosystem.

The Gulf Stream's current structure and eddy production became established during the middle and late Pleistocene, allowing an entire endemic molluscan fauna to evolve within the stable tropical conditions of the Lake Worth Lagoon and adjacent smaller lagoons in Martin County. Some of the most prominent of these endemic lagoonal species are listed here and are also illustrated in the Iconography at the end of this chapter. Unique marine ecosystems have survived within this pocket of warm tropical water and this area of special evolution was named the **Palm Beach Infraprovince** by Petuch and Berschauer (2021). There is nothing comparable anywhere else in the entire Carolinian Province, making it an area of special interest for marine conservationists. On the shallow mud flats and Turtle Grass beds, bivalves, such as the endemic venerid *Mercenaria hartae* (Figure 2.13) and the small endemic semelid *Semele donovani*

(Figure 2.14), are present in large beds and these are the main prey items for large predatory gastropods such as the endemic naticid, *Neverita (Glossaulax) delessertiana patriceae* (Figure 2.16), and the large endemic melongenid, *Melongena (Rexmela) corona winnerae* (Figures 2.17 and 2.18) (see Petuch and Myers, 2014). The oyster beds that are scattered along the lagoon shoreline house other endemic Lake Worth gastropods such as the predatory muricid, *Stramonita buchecki* (Figures 2.2 and 2.12) and the odd, elongated cerithiid, *Cerithium lutosum lindae* (Figure 2.14, which may prove to be a full, distinct species). The extensive Turtle Grass beds found in the Palm Beach and Martin County Lagoons also support large aggregations of the endemic modulid gastropod, *Trochomodulus calusa foxhalli* (Figure 2.15) and the tiny olivid *Olivella ellenae* (Figure 2.14 G, H). These, and other mollusks that are endemic to the Palm Beach Infraprovince, are listed here:

MOLLUSKS THAT ARE ENDEMIC TO THE PALM BEACH INFRAPROVINCE
Gastropoda
Cerithiidae
Cerithium lutosum lindae Petuch, 1987
Modulidae
Trochomodulus calusa foxhalli (Petuch and Myers, 2014)
Naticidae
Neverita (Glossaulax) delessertiana patriceae Petuch and Myers, 2014
Muricidae
Stramonita buchecki Petuch, 2013
Melongenidae
Melongena (Rexmela) corona winnerae Petuch, 2004
Olividae
Olivella ellenae Petuch and Berschauer, 2025
Bivalvia
Veneridae
Mercenaria hartae Petuch, 2013
Semelidae
Semele donovani McGinty, 1955

Along the Florida coast, north of the Palm Beach Infraprovince, in the area between Jupiter and Jacksonville, the coastal lagoons also house several interesting large endemic gastropods. Primary among these are two distinct subspecies of the western Floridian *Melongena (Rexmela) corona* (Gmelin, 1791): the High-Spired Crown Conch, *Melongena (Rexmela) corona altispira* Pilsbry and Vanatta, 1934 (Figure 2.19) and the Spruce Creek Crown Conch, *Melongena (Rexmela) corona sprucecreekensis* J. Tucker, 1994 (Figure 2.20). Since *Melongena (Rexmela)* species live only in sheltered, quiet-water lagoons with muddy seafloors, they cannot migrate into areas of open sand beaches. These high-energy oceanic wave environments act as ecological barriers between the three subspecies found along eastern Florida. The narrowing and attenuation of the coastal lagoon north of the Jupiter Inlet, along with the adjacent wide sand beaches, act as genetic barriers between the southern *Melongena (Rexmela) corona winnerae* and the central *M. (Rexmela) corona altispira*. The populations of the high-spired subspecies *altispira* are the largest in the Indian River, at Fort Pierce and Vero Beach, and within the Banana River lagoon system at Cape Canaveral. North of Cape Canaveral, the brackish Mosquito Lagoon and the wide sandy beaches near New Smyrna Beach act as genetic barriers between the central *M. (Rexmela) corona altispira* and the northern *M. corona sprucecreekensis*. This large subspecies ranges from near Daytona Beach northward

Figure 2.2 A Pair of Bucheck's Murex (*Stramonita buchecki*) laying eggs on a rock in Lake Worth Lagoon, northern Palm Beach County, Florida. This small muricid gastropod is known only from the Palm Beach Infraprovince area, where water conditions stay tropical even during the coldest winter weather. Photo courtesy of Robert Eason.

to the St. Augustine area and is the northernmost melongenid subspecies in the southern portion of the Georgian Subprovince.

These three Atlantic coast subspecies of *Melongena (Rexmela) corona* are only half of the entire species complex, as two other subspecies and one sibling species occur along the Gulf of Mexico coast of Florida (see Petuch and Berschauer, 2025 for illustrations of all the Gulf taxa and their many named ecophenotypical variants). The nominate subspecies, *Melongena (Rexmela) corona corona* (Gmelin, 1791), ranges from the Ten Thousand Islands of southwestern Florida northward to near Tampa. North of Tampa, the fluctuating salinities caused by the effluent of the Suwannee River, as well as the hyposaline conditions in Waccasassa Bay, act as the ecological barrier between the nominate subspecies and the giant northern subspecies, *Melongena (Rexmela) corona johnstonei* Clench and Turner, 1956. This large melongenid, which converges on the morphology of the northeastern Florida *M. (Rexmela) corona sprucecreekensis*, both in general shape and size, ranges from Cedar Key westward along the Florida Panhandle and into Mobile Bay, Alabama. The carbonate environments and coral reefs of the Florida Keys house the dwarf species *Melongena (Rexmela) bicolor* (Say, 1826) and its named ecophenotypic variants (*bicolor* may prove to be a subspecies of *corona*, but because of its great differences in shell morphology and ecology, Petuch and Berschauer, 2025 considered it to be a full sibling species). These coral environments also act as an ecological barrier between the eastern coast *Melongena (Rexmela) corona winnerae* and the western coast *M. (Rexmela) corona corona*, and the two subspecies do not overlap geographically. The dwarf Florida Keys *M. (Rexmela) bicolor* may have evolved during the late Pleistocene, when lowered sea levels and cold-water conditions decimated the local populations of *corona* along the emergent Florida coast. A small, isolated population of *corona* survivors found a refuge in the warmer water of what are now the Keys and evolved into a new sibling species. The entire *M. (Rexmela) corona* complex, with the exception of the Mobile Bay population of *johnstonei*, is restricted to Florida and is one of the most important groups of gastropods in the Carolinian Province ecosystems. As presently understood, this complex includes:

East Coast of Florida
Melongena (Rexmela) corona winnerae Petuch, 2004
Melongena (Rexmela) corona altispira Pilsbry and Vanatta, 1934
Melongena (Rexmela) corona sprucecreekensis J. Tucker, 1994
Florida Keys and Southern Biscayne Bay
Melongena (Rexmela) bicolor (Say, 1826)
West Coast of Florida
Melongena (Rexmela) corona corona (Gmelin, 1791)
Melongena (Rexmela) corona johnstonei Clench and Turner, 1956

The similar sibling species, *Melongena (Rexmela) bispinosa* (Philippi, 1844) and its variant *martiniana* (Philippi, 1844), are confined to the Yucatan Peninsula of Mexico (see Petuch and Berschauer, 2025). Because of adverse ecological conditions along the coasts of Texas, Louisiana, and Mississippi, no melongenids of any kind are present there.

Endemism in the Outer Neritic Zone Pectinid Beds

The single-most characteristic molluscan assemblage found within the Georgian Subprovince is one based on the widespread dense beds of the Carolina Calico Scallop, *Argopecten gibbus carolinensis* (Grau, 1952) (Figure 2.23). This large endemic subspecies of the smaller Calico Scallop (*Argopecten gibbus gibbus* from the Gulf of Mexico and Caribbean) occurs in immense, closely packed aggregations that cover large areas of the open outer neritic seafloor, from Cape Hatteras, North Carolina south to Cape Canaveral, Florida. Being over twice the size of the nominate subspecies, the Carolina scallop is also a much more inflated and globose shell that is not as colorful as its southern relative, having mostly dark red markings on the dorsal valve and lacking the intense orange, pink, and lavender colors that typify *A. gibbus gibbus*. Being a large and abundant scallop in the offshore areas along the edge of the continental shelf (Outer Neritic Zone), *A. gibbus carolinensis* forms the basis of a large commercial scallop harvesting business. The scallop boats that dredge these outer shelf waters in search of scallop beds also collect large numbers of other mollusks as by-catch in their nets. Most of the information on the biodiversity of the offshore molluscan faunas comes from the material brought up along with the scallops.

The Georgian Subprovince scallop beds house several characteristic molluscivorous gastropods, including the fasciolariids *Cinctura keatonorum* (Figure 2.27) and *Polygona williamlyonsi* (Figure 2.26) the volute *Scaphella sheltoni* (Figures 2.24 and 2.25), and the olivid *Americoliva bollingi clenchi* (Figure 2.28). The shell grit substrate upon which the scallops beds reside supports a large fauna of polychaete worms, and these are the principal prey items for vermivorous gastropods such as the conid *Gradiconus aliceae* (Figure 2.29) and the terebrid *Neoterebra onslowensis* (Figure 1.29, Chapter 1; also found in the Florida Keys and southwestern Florida, but most frequently encountered off the Carolinas). The scallop beds also harbor a distinctive dwarf form of the Small-Spined Helmet Shell, *Cassis spinella* (Figure 1.8, Chapter 1), which is especially common in the Onslow Bay and Cape Lookout areas of North Carolina. This dwarf population only grows to about half the length of *spinella* populations found elsewhere in the Carolinian Province. Some of the most prominent scallop bed-associated gastropods include:

ENDEMIC GASTROPODS THAT ARE ASSOCIATED WITH THE OFFSHORE CALICO SCALLOP BEDS

Gastropoda

Fasciolariidae

Cinctura keatonorum Petuch, 2013

Polygona williamlyonsi (Petuch and Sargent, 2011)

Volutidae

Scaphella sheltoni Petuch and Berschauer, 2025

Olividae

Americoliva bollingi clenchi Petuch and Berschauer, 2019

Cancellariidae

Axelella smithii (Dall, 1880)

Conidae

Gradiconus aliceae Petuch and Berschauer, 2025

Terebridae

Neoterebra onslowensis (Petuch, 1974)

Offshore Coral Bioherms of North Carolina

In 1969, two Duke University marine scientists, Drs. Ian Macintyre and Orrin Pilkey, published on their discovery of living Caribbean reef corals found off the Cape Hatteras and Cape Lookout areas. These coral bioherms had been discovered to be growing on raised rock knolls along the edge of the continental shelf and directly within the flow of the Gulf Stream. Unlike the rich coral fauna seen farther south in the Florida Keys, the North Carolina bioherms contained only two species of cold-tolerant star corals: *Siderastrea siderea* and *Solenastrea hyades* (Macintyre and Pilkey, 1969). On the offshore bioherms, these two corals take on an unusual morphology. Growing to sizes larger than normal individuals found farther south, both species take on a pyramidal shape with a projecting columnar structure at the top of the colony. Since Macintyre and Pilkey's original report, another large species of scleractinian coral, the star coral *Solenastrea bournoni*, has also been collected on the North Carolina bioherms.

Two years after this initial discovery, while participating in a research cruise on the Duke University oceanographic research vessel *R/V Eastward*, the senior author discovered a number of classic Caribbean gastropods living in association with these North Carolina coral assemblages (Petuch, 1972). Several species had never been reported from that far north, including Dennison's Morum, *Oniscidia dennisoni* (reported as *Morum dennisoni*), and the Milk Conch, *Macrostrombus costatus* (reported as *Strombus costatus*), and these were the subject of amazement by the marine biologists at the Duke University Marine Laboratory in Beaufort. The planktotrophic larvae of these two classic Caribbean gastropods were found to have been transported to the Hatteras coral bioherms by riding the Gulf Stream current, before it veers off into the open Atlantic. These two unexpected gastropods are shown here on Figures 2.30 and 2.32. Since that time, more research, including SCUBA diving and dredging on the coral bioherms, has been undertaken, and this has revealed many more previously unreported Caribbean species. Some of these are listed in Chapter 1 of this book and are illustrated here in the Iconography at the end of this chapter.

Figure 2.3 Map of the principal coral reefs of North Carolina (purple).

Deep Water Bathyal Faunas

At 200 m depth below surface (the outermost edge of the Neritic Zone), the broad continental shelf off the Carolinas, Georgia, and northeastern Florida plunges at a steep angle into the Bathyal Zone of the continental slope. This narrow strip of seafloor extends down to around 400 m depth and offers a number of unique habitats for a very rich fauna of mollusks. Since the rivers along the Eastern Seaboard exhibit Braided Stream depositional patterns, produced by rivers that flood during the spring and dry up during the winter, immense amounts of fine sediments are regularly deposited onto the eastern United States continental shelf. These fine sediments are the result of the continuous erosion of the slopes of the Appalachian Mountains by steep-gradient, high-energy Juvenile Streams. The sediments, mostly clays produced by the hydrolysis of feldspars and micas in granitic rocks, fan out onto the continental shelf and eventually cascade down the continental slope and into the Hatteras Abyssal Plain far offshore. In lowland South Carolina, these organic-rich sediments are referred to as "Pluff Muds" by the local inhabitants.

The deep Neritic Zone (100-200 m depth) area off South Carolina contains an interesting habitat that is essentially unique along the East Coast. Composed of huge blocks of compacted mud and clay and referred to as the "**Charleston Mud Lumps**", this rough seafloor was formed by massive accumulations of muds and clays produced by the effluent of the Waccamaw, Pee Dee, and Santee Rivers (all Braided Streams). This large array of high-relief, blocky submarine topography offers habitats for numerous rarely-seen deeper water species such as the gazid *Gaza superba*, the muricids *Murexiella hidalgoi*, *Pterochelus ariomus*, and *Babelomuex dalli*, the epitoniid *Stenorhytis pernobilis* (some shown here on Figure 2.37), the buccinids *Ptychosalpinx globulus* and *Liomesus stimpsoni*, the mitrid *Mitra antillensis* (all illustrated on Figure 2.38), the cone shells *Dalliconus mcgintyi* (Figure 1.25, Chapter 1) and *Conasprelloides villepinii fosteri* (Figure 2.35), and the cylichnid *Scaphander watsoni* (Figure 2.37). On muddy substrates at this same depth farther south, along the northeastern coast of Florida and the Georgia coast, the endemic scaphelline volute *Volutifusus aguayoi* (Figure 2.35) lives together with this same assemblage of gastropods. The small *"Scaphella" garciai*, named from a single specimen collected off St. Augustine, Florida, has now been found to be a dwarf form of *Volutifusus aguayoi* and is not considered to be a valid taxon. The widespread bathyal Gulf of Mexico scallop, *Euvola marshallae* (Figure 1.32, Chapter 1) has been dredged off the St. Augustine coast, showing that this paper scallop has a wider range than was originally thought (Petuch and Berschauer, 2025).

Farther offshore from the Mud Lump area of South Carolina, in the upper part of the continental slope (Bathyal Zone), lies another unusual series of habitats that are associated with the "**Charleston Bump**". This large, raised area represents a topographic high on the bathyal Blake Plateau and rises up to a depth of 375 m from the sea surface. Areas surrounding the Charleston Bump are composed of limestone of Eocene age and are riddled with large cave systems. The Gulf Stream and its associated countercurrents flow directly over the Charleston Bump, producing strong eddies and forming upwellings of cold, nutrient water from the adjacent Blake Plateau and Hatteras Abyssal Plain. The upwelling nutrients create rich plankton blooms, and these support entire communities of marine life, such as those associated with the large bioherms ("reefs") of the delicate branching ahermatypic coral *Lophelia*. Some of the more characteristic deep-water gastropods that live on the Charleston Bump, Blake Plateau, and lower continental slope include the abalone *Haliotis pourtalesi* (Figure 2.32), the pleurotomariids *Bayerotrochus charlestonensis* and *Perotrochus maureri* (both Figure 2.34), the fasciolariids *Harasewychia aepynotum* and *Aristofusus excavatus*, the turbinellid *Fulgurofusus timor*, the turrid *Cryptogemmula periscelida*, the cochlespirid *Cochlespira elegans*, and the epitoniid *Amaea retifera* (all shown on Figure 2.36). The Hatteras Abyssal Plain also houses endemic gastropods, including the large raphitomid *Bathybela costlowi* (Figure 2.38) (Petuch, 1974b), which erroneously had been considered to be a synonym of the eastern Atlantic *Bathybela tenelluna*, which is found off northwestern Africa and Portugal.

The following is a list of some of the more important gastropods and bivalves that are endemic to the Georgian Subprovince. Most of these are illustrated in the Iconography at the end of this chapter. Additional information and discussions are given on the captions of each figure.

GEORGIAN SUBPROVINCE ENDEMIC TAXA
Gastropoda
Pleurotomariidae
Bayerotrochus charlestonensis (Askew, 1988)
Perotrochus maureri Harasewych and Askew, 1993
Margaritidae
Margarites miona Dall, 1927 (very deep water)
Siliquariidae
Siliquaria squamata Blainville, 1827
Modulidae
Modulus pacei Petuch, 1987
Ovulidae
Pseudosimnia vanhyningi (M. Smith, 1940)
Muricidae
Eupleura caudata form *etterae* B. Baker, 1951
Pterochelus ariomus (Clench and Perez Farfante, 1945) (possibly the same species as *P. helenae* from Dominica)
Urosalpinx cinerea form *folleyensis* B. Baker, 1951
Fasciolariidae
Cinctura keatonorum Petuch, 2013
Fusinus thompsoni Hadorn and Rogers, 2000
Fusinus vitreus Dall, 1927
Polygona williamlyonsi (Petuch and Sargent, 2011)
Busyconidae
Busycon carica eliceans (Montfort, 1810)
Fulguropsis rachelcarsonae Petuch, Myers, and Berschauer, 2016
Sinistrofulgur sinistrum (Hollister, 1958) unnamed East Coast subspecies
Melongenidae
Melongena (Rexmela) corona altispira Pilsbry and Vanatta, 1934
Melongena (Rexmela) corona sprucecreekensis J.Tucker, 1994
Buccinidae
Colus perminutus (Dall, 1927)
Colus profundicola (Verrill and Smith, 1884)
Colus rushii (Dall, 1889)
Liomesus stimpsoni Dall, 188)
Mohnia carolinensis (Verrill, 1884)
Ptychosalpinx globulus (Dall, 1889)
Volutidae
Bathyaurinia aguayoi (Clench, 1940)
Bathyaurinia aguayoi dwarf form *garciai* (Bail, 2007) ("*Scaphella" garciai* Bail is a synonym)
Rehderia georgiana (Clench, 1946)
Scaphella sheltoni Petuch and Berschauer, 2025

Marginellidae
Prunum hartleyanum (Schwengel, 1941)
Prunum roscidum (Redfield, 1860)
Olividae
Americoliva bollingi clenchi Petuch and Berschauer, 2019
Americoliva sayana sayana (Ravenel, 1834)
Turbinellidae
Fulgurofusus timor Harasewych, 1983
Peristarium aurora (Bayer, 1971)
Cancellariidae
Axelella smithii (Dall, 1880)
Conidae
Gradiconus aliceae Petuch and Berschauer, 2025
Terebridae
Cinguloterebra floridana (Dall, 1889)
Neoterebra arcas (Abbott, 1954)
Neoterebra concava (Say, 1827)
Neoterebra onslowensis (Petuch, 1974)
Raphitomidae
Bathybela costlowi (Petuch, 1974)
Cylichni
Scaphander watsoni Dall, 1881
Bivalvia
Veneridae
Mercenaria mercenaria notata (Say, 1822)
Pectinidae
Argopecten gibbus carolinensis Grau, 1952
Argopecten irradians concentricus (Say, 1822)
Cardiidae
Dinocardium robustum (Lightfoot, 1786)

Iconography of Georgian Subprovince Mollusks

Figure 2.4 Sea Grass Periwinkle from the Cordgrass beds of the Georgian Subprovince and Virginian Province.

A, B= *Littoraria irrorata irrorata* (Say, 1822), length 19 mm; **C, D=** *Littoraria irrorata irrorata* (Say, 1822), length 21 mm; **E=** *Littoraria irrorata irrorata* (Say, 1822), length 20 mm. Restricted to the US East Coast, where it lives on *Spartina* Grass in intertidal areas from the Indian River area of eastern Florida northward to Long Island, New York. **F=** *Littoraria irrorata sayi* (Reeve, 1858), length 19 mm, Cedar Key, Florida. The genetically isolated Gulf subspecies *sayi* has finer and more numerous cords on the body whorl, a more inflated body whorl, a yellowish base color, and an orange columellar callus instead of a brown one like that of *irrorata irrorata*. Both subspecies are absent from the Florida Keys, where the coral reefs and carbonate environments act as an ecological barrier.

Figure 2.5 Say's Olive Shell from the sandy areas of the Georgian Subprovince.
A, B= *Americoliva sayana* (Ravenel, 1834), length 80 mm; **C, D**= *Americoliva sayana* (Ravenel, 1834), length 63 mm; **E, F**= *Americoliva sayana* (Ravenel, 1834), length 50 mm, a dark color form from Peanut Island, Lake Worth, Florida, at the southernmost end of the range of *A. sayana*. This classic Carolinian Province species has now been shown to comprise three separate geographically isolated subspecies: *Americoliva sayana sayana* which ranges from Cape Hatteras to Palm Beach County, Florida; *Americoliva sayana sarasotaensis* (Petuch and Sargent, 1986) which ranges from Pavilion Key, Ten Thousand Islands of southwestern Florida northward to the Florida Panhandle; and *Americoliva sayana texana* (Petuch and Sargent, 1986) which ranges from northern Texas south to Veracruz, Mexico. There are no *A. sayana* subspecies of any kind in the Florida Keys, where the coral reefs and carbonate environments act as an ecological barrier. The western Florida and Texas subspecies are illustrated and discussed in Petuch and Berschauer, 2025.

Figure 2.6 Concentric Bay Scallop from sea grass beds of the Georgian Subprovince.
A, B= *Argopecten irradians concentricus* (Say, 1822), length 55 mm; **C=** *Argopecten irradians concentricus* (Say, 1822), length 49 mm; **D=** *Argopecten irradians concentricus* (Say, 1822), length 52 mm. This classic Carolinian Province species has now been shown to comprise four separate geographically-isolated subspecies: *Argopecten irradians irradians* (Lamarck, 1819) which ranges from Cape Cod, Massachusetts south to New Jersey; *Argopecten irradians concentricus* (Say, 1822) which ranges from Cape Hatteras, North Carolina south to Fort Pierce, Florida; *Argopecten irradians taylorae* Petuch, 1987 which ranges from Biscayne Bay, Florida, throughout the Florida Keys, and northward along western Florida and the Florida Panhandle; and *Argopecten irradians amplicostatus* (Dall, 1898) which ranges from Texas south to the Yucatan Peninsula of Mexico. These subspecies, or possibly a complex of sibling species, live in sea grass beds within quiet sheltered lagoons all along the East Coast and throughout the Gulf of Mexico. The evolution and distribution of this species complex was described and discussed in Petuch and Berschauer, 2022.

Figure 2.7 Eastern Giant Cockle.

A, B= *Dinocardium robustum* (Lightfoot, 1786), length 74 mm; **C**= *Dinocardium robustum* (Lightfoot, 1786), length 63 mm; **D**= *Dinocardium robustum* (Lightfoot, 1786), length 52 mm. This large species ranges from Virginia southward to Fort Pierce, Florida. The genus *Dinocardium* is absent in the Florida Keys and southeastern Florida but is present along western Florida and throughout the entire Gulf of Mexico as the sister species, *Dinocardium vanhyningi* Clench and Smith, 1944. The differences between the two sibling species and their evolution were discussed in Petuch and Berschauer, 2022 and 2024.

Figure 2.8 Marginellas of the Georgian Subprovince.
A, B= *Prunum roscidum* (Redfield, 1860), length 11 mm, ranges from Massachusetts to eastern Florida; **C, D**= *Prunum canillum* (Dall, 1927), length 16 mm, a rare deep-water species that is found primarily off the coasts of Georgia and northeastern Florida; **E, F**= *Prunum carneum evelynae* (Bayer, 1943), length 22 mm, a northern subspecies of the Caribbean *Prunum carneum* (Storer, 1837) that is found from Cape Canaveral south to Palm Beach; **G, H**= *Prunum apicinum* (Menke, 1828), length 11 mm, a widespread and very common species found throughout the entire Carolinian Province.

Figure 2.9 Marked Venus Clam from the North Carolina coastal lagoons.
A, B, C= *Mercenaria mercenaria notata* (Say, 1822), length 106 mm; **D, E**= *Mercenaria mercenaria notata* (Say, 1822), length 43 mm. This ornately patterned subspecies of the northern Quahog, with its characteristic brown zig-zag color markings, is found in the quiet, protected water conditions within the coastal lagoons of the Carolinas. It is one of a complex of estuarine venerids that are found throughout the Carolinian Province, including species such as *Mercenaria hartae* Petuch, 2013 from southeastern Florida, *Mercenaria browni* Petuch and Berschauer, 2019 from the Ten Thousand Islands of southwestern Florida, and *Mercenaria texana* (Dall, 1902) from the Texas lagoons (see Petuch and Berschauer, 2025 for a review of this complex of estuarine venerids). Recently, *M. mercenaria notata* has been introduced to the northwestern coast of Florida, where large clam farms have been established (especially on Cedar Key). This Carolina subspecies escaped during hurricanes and has now become established in the eastern Gulf of Mexico; it may be hybridizing with the local *Mercenaria* species.

Figure 2.10 Rachel Carson's Pear Whelk.
A, B= *Fulguropsis rachelcarsonae* Petuch, Myers, and Berschauer, 2015, holotype, length 76 mm, trawled by commercial scallop fishermen form 20 m depth off Cape Canaveral, Brevard County, Florida; **C, D**= *Fulguropsis rachelcarsonae* Petuch, Myers, and Berschauer, 2015, length 96 mm, low tide on mud flats, Pine Point, Singer Island, Palm Beach County, Florida; **E, F**= *Fulguropsis rachelcarsonae* Petuch, Myers, and Berschauer, 2015, variant with keeled shoulder, length 70 mm, Jupiter Island, Palm Beach County, Florida. This heavily ribbed and highly sculptured whelk ranges along the entire southeastern coast of the United States, from Cape Hatteras, North Carolina to Palm Beach, Florida.

Figure 2.11 Giant Form of the Eastern Florida Alphabet Cone.

A, B= *Lindaconus atlanticus* (Clench, 1942), length 78 mm. Giant coastal lagoon variant, collected in Lake Worth Lagoon, under the Blue Heron Boulevard Bridge, Riviera Beach, Palm Beach County, Florida. This widespread Gulf of Mexico species also ranges across the Florida Keys and northward up the Florida East Coast to at least the Georgia border but may live offshore in deeper water farther north. The Palm Beach County lagoonal specimens attain the largest size of any known population.

Figure 2.12 Bucheck's Rock Shell from the Eastern Florida coastal lagoons.
A, B= *Stramonita buchecki* Petuch, 2013, length 30 mm; **C, D**= *Stramonita buchecki* Petuch, 2013, length 29 mm; **E, F**= *Stramonita buchecki* Petuch, 2013, length 26 mm. All specimens were collected at Pine Point, Singer Island, Lake Worth Lagoon, Palm Beach County, Florida. This small, recently discovered muricid gastropod lives on oyster beds growing within quiet, protected lagoons along the Florida East Coast.

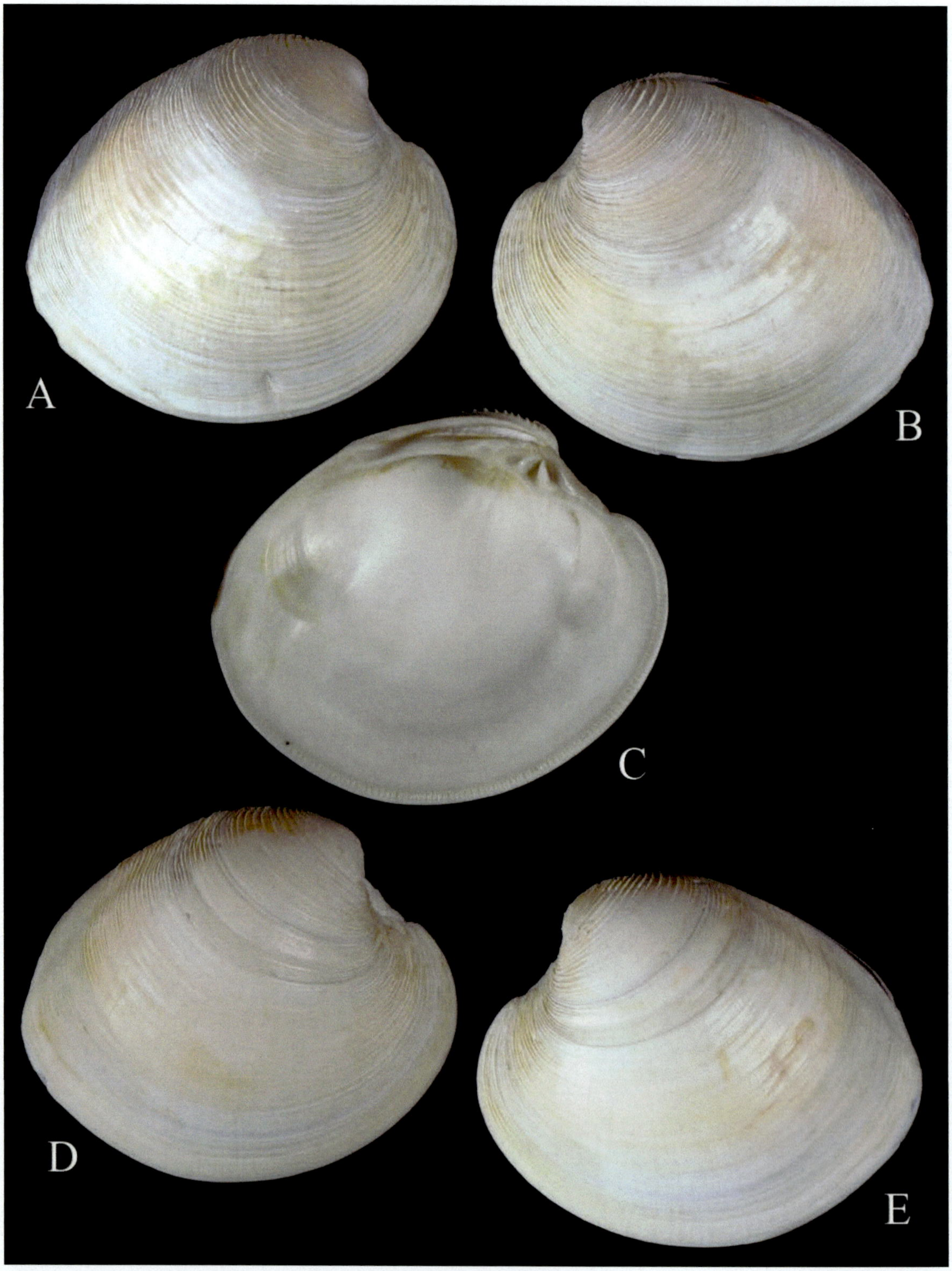

Figure 2.13 Hart's Venus Clam from the Eastern Florida coastal lagoons.
A, B= *Mercenaria hartae* Petuch, 2013, width 76 mm; **C=** *Mercenaria hartae* Petuch, 2013, width 78 mm; **D, E=** *Mercenaria hartae* Petuch, 2013, width 43 mm. All specimens were collected at Pine Point, Singer Island, Lake Worth Lagoon, Palm Beach County, Florida. This smooth-sculptured, pale-yellow venerid is only known from the lagoons along southeastern Florida, where it ranges from Fort Pierce south to Lake Worth.

Figure 2.14 Endemic mollusks of the Eastern Florida coastal lagoons.
A, B= *Semele donovani* McGinty, 1955, length 13 mm; **C, D=** *Cerithium lutosum lindae* Petuch, 1987, length 12 mm; **E, F=** *Cerithium lutosum lindae* Petuch, 1987, length 12 mm; **G, H=** *Olivella ellenae* Petuch and Berschauer, 2025, length 11 mm. All specimens were collected at Pine Point, Singer Island, Lake Worth Lagoon, Palm Beach County, Florida. The very elongated *Cerithium lutosum lindae* may prove to be a full species all to itself and distinct from the widespread *C. lutosum*. These endemic Florida coastal gastropods range as far northward as Cape Canaveral.

Figure 2.15 Endemic Modulid Gastropods from the Eastern Florida coastal lagoons.
A, B= *Modulus pacei* Petuch, 1987, width 14.5 mm; **C, D=** *Modulus pacei* Petuch, 1987, width 15 mm; **E=** *Modulus pacei* Petuch, 1987, width 16 mm; **F, G=** *Trochomodulus calusa foxhalli* Petuch and Myers, 2014, width 12 mm. All specimens were collected at Pine Point, Singer Island, Lake Worth Lagoon, Palm Beach County, Florida. The subspecies *foxhalli* differs from the Florida Keys endemic *Trochomodulus calusa* (Petuch, 1987) in having a higher spire and more trochoidal shape, and in having more numerous and stronger spiral cords on the shell base and spire whorls. Both *Modulus pacei* and *Trochomodulus calusa foxhalli* range as far north as Cape Canaveral.

Figure 2.16 Patrice's Moon Snail from the Eastern Florida coastal lagoons.
A, B, C= *Neverita (Glossaulax) delessertiana patriceae* Petuch and Myers, 2014, height 52 mm; **D, E**= *Neverita (Glossaulax) delessertiana patriceae* Petuch and Myers, 2014, height 56 mm. This East Coast subspecies of the Gulf of Mexico *Neverita (Glossaulax) delessertiana* is found in shallow coastal lagoons from Palm Beach County, Florida north to Cape Canaveral, Florida. The subspecies grows to its largest size in Lake Worth Lagoon, Palm Beach County.

Figure 2.17 Winner's Crown Conch.
A, B= *Melongena (Rexmela) corona winnerae* Petuch, 2004, length 128 mm; **C, D=** *Melongena (Rexmela) corona winnerae* Petuch, 2004, length 126 mm. This large, inflated subspecies is found in coastal lagoons from Jupiter, Palm Beach County, Florida south to Lake Worth, Palm Beach County, Florida and grows to its maximum size in Lake Worth Lagoon.

Figure 2.18 Giant Form of Winner's Crown Conch.
A, B= *Melongena (Rexmela) corona winnerae* Petuch, 2004, giant form, length 149 mm. Found on a sandy mudflat at low tide, Pine Point, Singer Island, Little Lake Worth Lagoon, Palm Beach County, Florida. Giant specimens like this one often have high, protracted spire whorls, approaching those seen on the smaller subspecies *altispira.*

Figure 2.19 High-Spired Crown Conch from the Central-Eastern Florida coastal lagoons.
A, B= *Melongena (Rexmela) corona altispira* Pilsbry and Vanatta, 1934, length 67 mm; **C, D=** *Melongena (Rexmela) corona altispira* Pilsbry and Vanatta, 1934, length 71 mm; **E, F=** *Melongena (Rexmela) corona altispira* Pilsbry and Vanatta, 1934, length 69 mm. Found on oyster banks at low tide, at Micco, Brevard County, Florida. Confined to the east-central coast of Florida, from Daytona Beach, Volusia County, south to Stuart, Martin County.

Figure 2.20 Spruce Creek Crown Conch.
A, B= *Melongena (Rexmela) corona sprucecreekensis* Tucker, 1994, length 140 mm; **C, D=** *Melongena (Rexmela) corona sprucecreekensis* Tucker, 1994, length 128 mm. Collected on oyster banks at low tide, in Spruce Creek, Volusia County, Florida. This is the farthest-north *Melongena (Rexmela)* subspecies found along the eastern coast of the United States.

Figure 2.21 Ornate Lightning Whelk from the Carolina coastal lagoons.
A, B= *Busycon carica eliceans* (Montfort, 1810), length 136 mm, in a tidal creek on Sapelo Island, McIntosh County, Georgia; **C, D=** *Busycon carica eliceans* (Montfort, 1810), length 115 mm, on sand flats off Edisto Island, Colleton County, South Carolina. This inflated subspecies may eventually prove to be a distinct full species.

Figure 2.22 Southern Form of the Snow Whelk.
A, B= *Sinistrofulgur laeostomum* (Kent, 1982), length 195 mm, found on the beach after a strong storm, Nags Head, Dare County, North Carolina. This large, knobbed form ranges southward to northeastern Florida.

Figure 2.23 Carolina Calico Scallop.

A, B= *Argopecten gibbus carolinensis* Grau, 1952, dorsal and ventral valves, length 53 mm; **C, D**= *Argopecten gibbus carolinensis* Grau, 1952, dorsal and ventral valves, length 52 mm; **E**= *Argopecten gibbus carolinensis* Grau, 1952, dorsal valve, length 57 mm. All three specimens were dredged by commercial scallop boats from 25 m depth off Beaufort, Carteret County, North Carolina. This distinctive subspecies of the shallow-water *Argopecten gibbus gibbus* (Linnaeus, 1758) differs in being a much larger and far more inflated shell, with extremely rounded valves that produce an almost globular shell profile. These shell characters are in stark contrast to those of the consistently smaller nominate subspecies, which has a much more flattened shell profile with fewer ribs. The subspecies *carolinensis* ranges from Cape Hatteras, North Carolina south to Fort Pierce, Florida and forms immense shoals of individuals in deeper water offshore.

Figure 2.24 Shelton's Volute from the offshore scallop beds.
A, B= *Scaphella sheltoni* Petuch and Berschauer, 2025, holotype, length 131 mm; **C, D**= *Scaphella sheltoni* Petuch and Berschauer, 2025, length 104 mm; **E, F**= *Scaphella sheltoni* Petuch and Berschauer, 2025, length 81 mm. This distinctive scaphelline volute is restricted to the offshore *Argopecten gibbus carolinensis* beds from Cape Hatteras, North Carolina, south to Cape Canaveral, Florida. *Scaphella sheltoni* differs from the well-known species, *Scaphella junonia* from western Florida, in consistently having thinner, more bladelike, and less developed columellar plicae and more numerous ribs on the early whorls.

Figure 2.25 Golden Form of Shelton's Volute.
A, B= *Scaphella sheltoni* Petuch and Berschauer, 2025, length 110 mm, a rare dark golden color form trawled by scallop fishermen north of Cape Canaveral, Florida.

Figure 2.26 Lyons' Latirus from the offshore scallop beds.
A, B= *Polygona williamlyonsi* Petuch and Sargent, 2011, holotype (USNM 1152531, Smithsonian Institution, Washington, D.C.), length 45.8 mm, dredged by commercial scallop boats from 50 m depth off Fort Pierce, St. Lucie County, Florida. This distinctive fasciolariid is restricted to the offshore *Argopecten gibbus carolinensis* scallop beds that extend from Cape Hatteras, North Carolina, south to off Fort Pierce, Florida. It is a rarely seen species with only a few specimens having ever been collected. The species was originally described in the Systematic Appendix in the back of Petuch & Sargent, 2011.

Figure 2.27 Keaton's Banded Tulip from the offshore scallop beds.
A, B= *Cinctura keatonorum* Petuch, 2013, length 103 mm; **C, D=** *Cinctura keatonorum* Petuch, 2013, length 92 mm; **E=** *Cinctura keatonorum* Petuch, 2013, length 94 mm. This distinctive fasciolariid is restricted to the offshore *Argopecten gibbus carolinensis* scallop beds that extend from Cape Hatteras, North Carolina, south to Fort Pierce, Florida.

Figure 2.28 Clench's Olive Shell from the offshore scallop beds.
A, B= *Americoliva bollingi clenchi* Petuch and Berschauer, 2019, length 53 mm; **C, D=** *Americoliva bollingi clenchi* Petuch and Berschauer, 2019, length 50 mm; **E, F=** *Americoliva bollingi clenchi* Petuch and Berschauer, 2019, length 48 mm. This distinctive, yellow-tinged olive shell lives on, and is restricted to, the offshore *Argopecten gibbus carolinensis* scallop beds that extend from Cape Hatteras, North Carolina south to Fort Pierce, Florida. Clench's Olive was originally described as a subspecies of the Caribbean Province *Americoliva nivosa* (Marrat, 1871) but is now known to be a northern subspecies of the Florida Keys endemic, *Americoliva bollingi* (Clench, 1934). All three specimens were dredged by commercial scallop boats, from 30-40 m depths off Mayport, Duval County, Florida.

Figure 2.29 Alice's Cone Shell.

A, B= *Gradiconus aliceae* Petuch and Berschauer, 2025, length 56 mm, trawled by commercial scallop boats from 30 m depth north of Cape Canaveral, Brevard County, Florida; **C, D=** *Gradiconus aliceae* Petuch and Berschauer, 2025, length 34 mm, dredged by commercial scallop boats from 30 m depth off Beaufort, Carteret County, North Carolina; **E, F=** *Gradiconus aliceae* Petuch and Berschauer, 2025, holotype (USNM 83854) length 37.5 mm, trawled by commercial scallop boats from 30 m depth off Cape Hatteras, North Carolina. This distinctive cone shell is restricted to offshore areas from Cape Hatteras, North Carolina south to Fort Pierce, Florida, where it is associated with the offshore scallop beds. This Georgian Subprovince shell has been incorrectly referred to as "*Conus largillierti* Kiener, 1847" in several recent publications (such as Mosher, 2024). That name, however, refers to a similar but different cone that is endemic to the western and northern Gulf of Mexico (see Petuch and Berschauer, 2025). The type locality for Kiener's taxon is the "Cotes du Mexique" ("Coast of Mexico"), demonstrating that *largillierti* was named from a Gulf of Mexico specimen and is not the same as the isolated East Coast species.

Figure 2.30 Milk Conch from the North Carolina Coral Bioherms.
A, B= *Macrostrombus costatus* (Gmelin, 1791), length 142 mm. This widespread large tropical strombid was first discovered to be living off North Carolina in the early 1970s (see Petuch, 1972). There it lives in association with the coral bioherms found off Cape Hatteras and Cape Lookout.

Figure 2.31 Cowries of the North Carolina Coral Bioherms.
A, B= *Naria acicularis* (Gmelin, 1791), length 24 mm; **C, D=** *Luria cinerea* (Gmelin, 1791), length 24 mm; **E, F=** *Macrocypraea zebra* (Linnaeus, 1758), length 78 mm. Found throughout the entire Carolinian Province, from Cape Hatteras, North Carolina south to the Florida Keys and the entire Gulf of Mexico, and also throughout the entire Caribbean Province. These cowries are frequently found on the coral bioherms off Cape Hatteras and Cape Lookout.

Figure 2.32 Mollusks of the North Carolina Coral Bioherms.
A, B= *Oniscidia dennisoni* (Reeve, 1843), length 36 mm; **C=** *Babelomurex scalariformis* (Lamarck, 1822), length 25 mm; **D=** *Babelomurex lintoni* (Verrill, 1882), length 32 mm; **E, F=** *Haliotis pourtalesii* Dall, 1881, length 12 mm (deep reefs and *Lophelia* coral bioherms); **G=** *Cirsotrema (Rectacirsa) dalli* McGinty, 1940, length 29 mm; **H=** *Calliostoma pulchrum* (C. Adams, 1850), height 14 mm. These wide-ranging western Atlantic species have been collected on the coral bioherms off Cape Hatteras, showing that they are also present in the farthest northern parts of the Georgian Subprovince. The widespread western Atlantic harpid, *Oniscidia dennisoni*, was first discovered to be living off North Carolina in the early 1970s (see Petuch, 1972). There it lives in association with the coral bioherms found off Cape Hatteras and Cape Lookout.

Figure 2.33 Mollusks of the North Carolina Coral Bioherms.
A, B= *Favartia goldbergi* Petuch and Sargent, 2011, length 11 mm; **C=** *Gemophos filistriatus* Vermeij, 2006, length 20.4 mm; **D, E=** *Cirsotrema (Rectacirsa) pilsbryi* Rehder, 1945, length 11 mm; **F, G=** *Lampasopsis thomae* (d'Orbigny, 1842), length 26 mm; **H=** *Calliostoma psyche* Dall, 1889, width 23 mm. These widespread Georgian Subprovince species have been collected on the coral bioherms off Cape Hatteras and Cape Lookout.

Figure 2.34 Pleurotomariid Gastropods of the Georgian Subprovince.
A, B= *Perotrochus maureri* Harasewych and Askew, 1993, holotype, width 60 mm; **C, D**= *Bayerotrochus charlestonensis* Askew, 1988, holotype, width 87 mm. These eastern United States slit shells live in deep water (200-400 m depth) along the Bathyal Zone, from South Carolina to Florida.

Figure 2.35 Deep-Water Gastropods of the Georgian Subprovince.
A, B= *Volutifusus aguayoi* form *garciai* (Bail, 2007), holotype, length 47.2 mm, dredged from 786 m depth off St. Augustine, St. Johns County, Florida (this small volute appears to be only a dwarf form of *V. aguayoi*); **C, D=** *Volutifusus aguayoi* (Clench, 1940), holotype, length 113 mm, dredged from 750 m depth off St. Augustine, St. Johns County, Florida (taken from Weaver and DuPont, 1970); **E, F=** *Conasprelloides villepinii fosteri* (Clench and Aguayo, 1942), length 53 mm; **G=** *Conasprelloides villepinii fosteri* (Clench and Aguayo, 1942), length 52 mm; **H=** *Conasprelloides villepinii fosteri* (Clench and Aguayo, 1942), length 62 mm. *Volutifusus aguayoi* is only known from very deep water off northeastern Florida and Georgia. *Conasprelloides villepinii fosteri*, which ranges from Cuba to Cape Hatteras, is found primarily within the Georgian Subprovince. It differs from the widespread Caribbean nominate subspecies, *Conasprelloides villepini villepinii* (Fischer and Bernardi, 1859), in having a much more elongated shell and a distinctly lower spire.

Figure 2.36 Deep-Water Gastropods of the Georgian Subprovince.
A= *Fulgurofusus (Peristerium) timor* Harasewych, 1983, holotype, length 52 mm; **B=** *Amaea retifera* (Dall, 1889), length 20 mm; **C=** *Cochlespira elegans* (Dall, 1881), length 34 mm (found in the Florida Straits); **D=** *Harasewychia aepynota* (Dall, 1889), length 19 mm (found in the Florida Straits and off Georgia); **E=** *Cryptogemmula periscelida* (Dall, 1889), length 43 mm; **F=** *Aristofusus excavatus* (Sowerby II, 1880), length 64 mm; **G, H=** *Pseudosimnia vanhyningi* (M. Smith, 1940), length 7 mm, on deep-water antipatharian octocorals; **I=** *Ventrilia tenera* (Philippi, 1848), length 24 mm. *Fulgurofusus timor* is known only from deep water off the Carolinas and was named for Cape Fear; from the Latin "timor", meaning "fear". All the other species shown here are classic deep-water gastropods that are found throughout the Carolinian Province.

Figure 2.37 Deep-Water Gastropods of the Georgian Subprovince.
A= *Scaphander watsoni* Dall, 1881, length 18 mm; **B, C=** *Gaza superba* (Dall, 1881), width 32 mm; **D, E=** *Stenorhytis pernobilis* (Fischer and Bernardi, 1857), length 45 mm; **F, G=** *Murexiella hidalgoi* (Crosse, 1869), length 39 mm; **H=** *Babelomurex dalli* (Emerson and d'Attilio, 1963), length 27 mm. These are all classic deep-water gastropods that are found throughout both the Carolinian and Caribbean Provinces.

Figure 2.38 Deep-Water Gastropods of the Georgian Subprovince.
A, B= *Ptychosalpinx globulus* (Dall, 1889), length 42 mm; **C, D=** *Liomesus stimpsoni* Dall, 1889, length 34 mm; **E=** *Mitra antillensis* Dall, 1889, length 62 mm; **F, G=** *Architectonica sunderlandi* Petuch, 1987, holotype, width 18 mm, recently incorrectly synonymized with the noticeably different cold-water South Atlantic species, *Adelphotectonica uruguaya* (Carcelles, 1953); that species has a distinctly rounded periphery and much more inflated body whorl, quite unlike the sharp-edged periphery and flattened body whorl of the deep-water Florida Straits and Florida Keys *Architectonica sunderlandi* (see Bieler and Petit, 2005); **H, I=** *Bathybela costlowi* (Petuch, 1974), holotype, length 28 mm, from 1000 m depth on the Hatteras Abyssal Plain (originally described as a *Pleurotomella*). This large North Carolina raphitomid has been incorrectly synonymized with *Bathybela tenelluna* Locard, 1897, a species found only along northwestern Africa and southernmost Portugal.

View of an exposure of the late Pleistocene (Tarantian Age) Anastasia Formation at Jupiter, Palm Beach County, Florida. At this location, known as "Blowing Rocks", the exposed limestone outcrops are eroded by the relentless wave action and contain many large holes and interconnected caverns. At high tide, water flowing into these "blow holes" creates fountains of spray that shoot high into the air, producing a geological feature that is unique to Palm Beach County.

CHAPTER 3. The Virginian Province

The Virginian Province is a geographically small area that represents a transition zone between the subtropical marine climate of the Carolinian Province to the south, and the cold subarctic water of the Acadian Province to the north. Named for the Commonwealth of Virginia, this transitional bioregion extends from Cape Hatteras, North Carolina northward to Cape Cod, Massachusetts (Figure 3.1). In stark contrast to the Georgian Subprovince (Chapter 2 of this book), the molluscan fauna of the Virginian Province is very impoverished, containing only a few hundred species in total. The low level of biodiversity seen today is the result of climatic degeneration during the late Pleistocene, which resulted in ecological collapse and the mass extinction of an older and richer molluscan fauna (see Chapter 5 of this book). The malacofauna of the central region of the Virginian Province, near the New York City area, was described by Jacobson and Emerson (1971).

Judging from the Miocene and Pliocene fossil record of Virginia, Maryland, Delaware, and New Jersey, the predecessor faunas of the Virginian Province contained large numbers of tropical, genera, such the gastropods *Eudolium* (Tonnidae), *Xenophora* (Xenophoridae), *Americoliva* (Olividae), and *Gradiconus* (Conidae) (discussed and illustrated in Chapters 6 and 7). All of these warm-water taxa disappeared from these states during the late Pleistocene and are now extant only south of Cape Hatteras, in the Carolinian Province. This abrupt faunal impoverishment can be traced directly to the build-up of massive continental glaciers, many extending as far south as northern New Jersey. These late Pleistocene glacial times, especially during the formation of the immense Laurentide Ice Sheet, brought high arctic marine conditions all the way south to Cape Hatteras and obliterated the resident neritic ecosystems. As a testament to the severity of the cold-water conditions during the Laurentide Glaciation, giant gouges ("keels") have been discovered on the continental shelf off Cape May, New Jersey, the result of immense icebergs being dragged along the shallow sea-floor by currents as they drifted southward (Duncan and Goff, 2001). The extinct subarctic gastropod, *Neptunea (Atractodon) stonei* (Figure 8.7, Chapter 8), collected in fossil beds from Nova Scotia to Cape Hatteras, lived among these ice bergs and boreal sea temperatures, in an environment very similar to Greenland today. Because of its preference for freezing, arctic water conditions, the presence of Stone's Neptune Shell can be used as a marker for the southernmost extent of the coldest marine climate of the late Pleistocene.

During the latest Pleistocene (around 25,000 years BP), the Laurentide Ice Sheet began to retreat, water temperatures warmed, and the sea level rose to almost present-day level. At this time, the newly flooded coastline and its warmer marine climate provided empty ecological niches that were rapidly occupied by colonizing species from the Carolinian Province. During the coldest times of the late Pleistocene, the Georgian Subprovince acted as a refuge (ecological refugium) for a number of wide-ranging warm-temperate genera, such as the gastropods *Neverita (Glossaulax)* (Naticidae), *Urosalpinx* and *Eupleura* (both Muricidae), *Busycon*, *Sinistrofulgur*, and *Busycotypus* (all Busyconidae), and *Ilyanassa* (Nassariidae), and the bivalves *Mercenaria* (Veneridae) and *Argopecten* (Pectinidae). All of these genera occur in the Carolinian Province of today and also in the Pliocene and Pleistocene fossil beds from Florida to Maryland. Examples of these physiologically hardy mollusks are shown here in the Iconography at the end of this chapter.

DOI: 10.1201/9781003732983-3

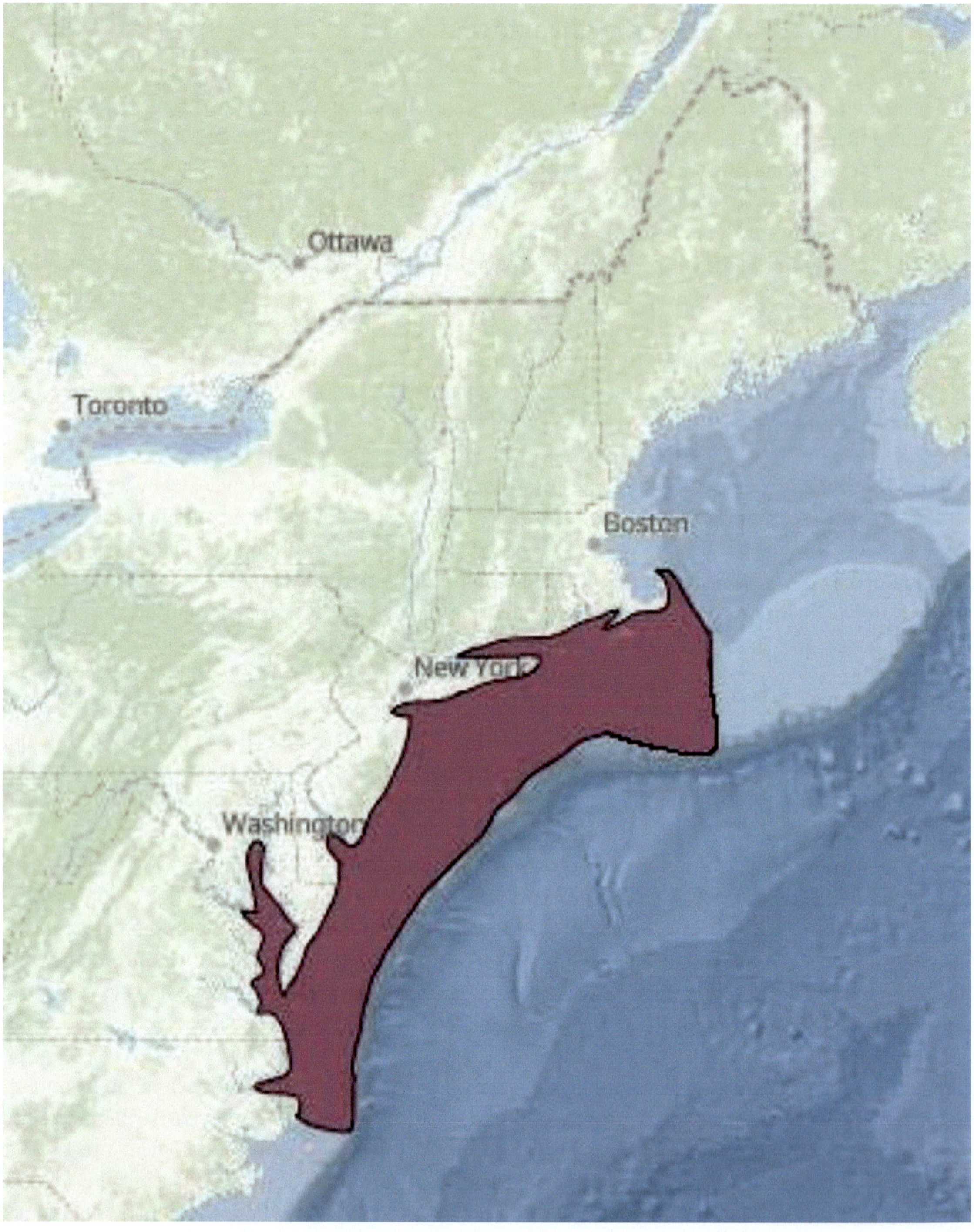

Figure 3.1 Map of the Virginian Molluscan Province. This is the only eastern American bioregion that contains but a single endemic mollusk, the scallop *Argopecten irradians irradians*. All the other faunal components are wide-ranging species, found in either the Acadian or Carolinian Provinces or in both. For this reason, the "province" actually represents a broad area of faunal transition between two well-defined provinces and does not conform to the 50% Rule of provinciality.

Unlike the Carolinian Province, with five distinct subprovinces, the Virginian Province houses a uniform molluscan fauna, with no major compositional differences anywhere within the bioregion and no definable subprovinces. The only exception is the presence of the large Snow Whelk, *Sinistrofulgur laeostomum* (Figure 3.9), which is found only within the southern part of the Virginian Province, from southern New Jersey to Cape Hatteras. This large left-handed species lives along with two other large busyconids, *Busycon carica* and *Busycotypus canaliculatum*, and all three whelk species range down to Cape Canaveral, Florida. From New Jersey to Florida, these busyconids are the primary predators within shallow coastal lagoons and the inner continental shelf, feeding on large bivalves such as *Crassostrea virginica* (Ostreidae), *Mercenaria mercenaria* (Veneridae), and *Spisula solidissima* (Mactridae). Along the New Jersey coast, all three busyconid species are harvested commercially and form the foundation of a fisheries industry that supplies the ethnic culinary market. Meat from the three busyconid species is the basis for the various *scungilli* dishes that are the highlights of any traditional Italian sea food dinner.

At the southern end of the Virginian Province, along Chincoteague Island and the Delmarva Peninsula coast of Virginia, a small pocket of genetic isolation and incipient speciation seems to be present in the shallow intertidal areas. Here, in the Shoal Grass and Eel Grass beds, the cerithiid *Bittiolum alternatum*, normally a tiny shell, grows to a very large size, often exceeding 20 mm in length. This giant form, that is unique to the Chincoteague Island area, was given the form or subspecies name *virginicum* by Henderson and Bartsch in 1914 (shown here on Figure 3.6). The oyster (*Crassostrea virginica*) banks around Chincoteague Island also house two unusual muricid gastropods, both giant variants of normally small species. These oversized ecological forms include *Urosalpinx cinerea* form *folleyensis* (Figure 3.4) and *Eupleura caudata* form *etterae*, and both are major predators on the local oysters. The presence of three different gigantic variants of small species in one limited area points to the presence of unique ecological parameters. The wide-ranging warm-water terebrid, *Neoterebra dislocata*, is sometimes found along the northern side of Cape Hatteras and along the Virginia coast, demonstrating that the oceanographic conditions in this area differ from those that occur farther north along the Delmarva Peninsula.

The Virginian Province is unique in that it contains only one endemic taxon; the bay scallop *Argopecten irradians irradians*. All the other resident mollusks have ranges that extend either into the Acadian Province to the north or into the Carolinian Province to the south, or both. For this reason, this "province" does not actually conform to the "50% Rule" of provinciality (Petuch, 2013; Petuch and Berschauer, 2020) and is simply a broad faunal transition zone between two distinct, quantitatively definable molluscan provinces. Because of its large geographical size, we here refer to this zone as a "province" only to retain a consistency in the biogeographic nomenclature. The following is a list of some of the more important gastropods and bivalves that are found in the faunally impoverished Virginian Province (Virginian Transition Zone). Most of these are illustrated in the Iconography at the end of this chapter. Additional information and discussions are given in the captions of each figure.

SOME CHARACTERISTIC MOLLUSKS OF THE VIRGINIAN PROVINCE

Gastropoda

Littorinidae

Littoraria irrorata irrorata (Say, 1822)

Cerithiidae

Bittiolum alternatum (Say, 1822)

Bittiolum alternatum form *virginicum* (Henderson and Bartsch, 1914)

Crepidulidae

Crepidula fornicata (Linnaeus, 1758)

Crepidula plana Say, 1822

Naticidae

Neverita (Glossaulax) duplicata (Say, 1822)

Muricidae

Urosalpinx cinerea (Say, 1822)

Urosalpinx cinerea form *folleyensis* B. Baker, 1851

Eupleura caudata (Say, 1822)

Eupleura caudata form *etterae* B. Baker, 1951

Busyconidae

Busycon carica (Gmelin, 1791)

Busycotypus canaliculatus (Linnaeus, 1758)

Sinistrofulgur laeostomum (Kent, 1982)

Nassariidae

Ilyanassa obsoleta (Say, 1822)

Hinia trivittata (Say, 1822)

Epitoniidae

Epitonium (Pictoscala) rupicola (Kurtz, 1860)

Melampidae

Melampus bidentatus Say, 1822

Bivalvia

Pectinidae

Argopecten irradians irradians (Lamarck, 1819) (confined to the Virginian Province; the only endemic taxon)

Ostreidae

Crassostrea virginica (Gmelin, 1791)

Veneridae

Mercenaria mercenaria (Linnaeus, 1758)

Petricolaria pholadiformis (Lamarck, 1818)

Mactridae

Spisula solidissima (Dillwyn, 1817)

Phariidae

Ensis directus (Conrad, 1844)

Iconography of Virginian Province Index Species

Figure 3.2 Common Slipper Shell.
A, B= *Crepidula fornicata* (Linnaeus, 1758), length 24 mm; **C, D=** *Crepidula fornicata* (Linnaeus, 1758), length 20 mm; **E, F=** *Crepidula fornicata* (Linnaeus, 1758), length 30 mm. This widespread Carolinian gastropod ranges from Labrador, Canada to the Florida Keys and throughout the Gulf of Mexico. It was recently introduced to northern European waters as escapees from mariculture farms that were raising imported American oysters. It has now become a pest organism in the estuaries of the Netherlands and Belgium.

Figure 3.3 Northern Shark Eye Moon Snail.
A, B, C= *Neverita (Glossaulax) duplicata* (Say, 1822), length 44 mm, Cape Hatteras, NC; **D, E=** *Neverita (Glossaulax) duplicata* (Say, 1822), length 53 mm, Lynn Beach, MA. This classic eastern United States shallow-water gastropod ranges from Cape Cod, Massachusetts south to Fort Pierce, Florida. The species is absent from the Florida Keys, and the coral reefs and carbonate environments there act as ecological barriers between the East Coast nominate subspecies and its two Gulf of Mexico subspecies: *Neverita (Glossaulax) duplicata campechiensis* (Recluz, 1843) in the eastern and northern Gulf of Mexico, and *Neverita (Glossaulax) duplicata texana* (Philippi, 1849) along the coast of Texas in the western Gulf of Mexico. These two subspecies are discussed and illustrated in Petuch and Berschauer, 2025. Specimens of *N. (Glossaulax) duplicata duplicata* from the northern part of its range have higher spires than those from the southern part of the range (Jacobson and Emerson, 1961).

Figure 3.4 Oyster Drill Muricids of the Virginian Province.
A, B= *Urosalpinx cinerea* (Say, 1822), length 25 mm; **C, D=** *Urosalpinx cinerea* (Say, 1822), length 24 mm; **E, F=** *Urosalpinx cinerea* (Say, 1822), length 21 mm.; **G=** *Urosalpinx cinerea* (Say, 1822) giant form *folleyensis* B. Baker, 1951, length 42 mm; an ecological variant (ecophenotype) found only on Chincoteague Island, Virginia, off the Delmarva Peninsula. *Urosalpinx cinerea* has been accidentally introduced into the northeastern Gulf of Mexico by clam farms using North Carolina stock. Although considered to be a classic Virginian Province index species, this hardy species also lives in the Acadian and Carolinian Provinces and ranges from Nova Scotia to central-eastern Florida.

Figure 3.5 Thick-Lipped Oyster Drill.
A, B= *Eupleura caudata* (Say, 1822), length 28 mm; **C, D=** *Eupleura caudata* (Say, 1822), length 24 mm; **E, F=** *Eupleura caudata* (Say, 1822), length 21 mm; **G=** *Eupleura caudata* giant ecological form *etterae* B. Baker, 1951, length 35.5 mm, found only on Chincoteague Island, Virginia, off the Delmarva Peninsula. This small muricid is a classic Virginian Province index species, also found throughout the Georgian Subprovince of the Carolinian Province, ranging from Cape Cod south to Palm Beach County, Florida.

Figure 3.6 Mud Snails and Dwarf Ceriths.
A, B= *Ilyanassa obsoleta* (Say, 1822), length 18 mm; **C=** *Ilyanassa obsoleta* (Say, 1822), length 20 mm; **D, E=** *Hinia trivittata* (Say, 1822), length 20 mm; **F=** *Hinia trivittata* (Say, 1822), length 12 mm; **G, H=** *Bittiolum alternatum* form *virginicum* (Henderson and Bartsch, 1914), length 18 mm (found only along Chincoteague Island, Virginia and lagoons north of Cape Hatteras). The *Ilyanassa* and *Hinia* species are found from Labrador south to Jacksonville, Florida and inhabit the Acadian, Virginian, and Carolinian Molluscan Provinces, often occurring in immense aggregations on exposed mud flats.

Figure 3.7 Lightning Whelk.
A, B= *Busycon carica* (Gmelin. 1791), length 224 mm. This large species ranges from Cape Cod, Massachusetts south to northeastern Florida and is found in both the Virginian and Carolinian Provinces. Exceptional specimens, with a bright-red aperture like the one shown here, are found along southern New Jersey, Delaware Bay, and the Cape Hatteras region of North Carolina. The specimen shown here was collected on sand flats at low tide, off Shackleford Island, Carteret County, North Carolina.

Figure 3.8 Channeled Whelk.
A, B= *Busycotypus canaliculata* (Linnaeus, 1758), length 168 mm. This large species ranges from Cape Cod, Massachusetts south to northeastern Florida and is found in both the Virginian and Carolinian Provinces. Channeled whelks are harvested off New Jersey and New York as one of the sources of the Italian molluscan seafood delicacy, *scungilli*.

Figure 3.9 Northern Form of the Snow Whelk.
A, B= *Sinistrofulgur laeostomum* (Kent, 1982), length 178 mm, dredged by commercial whelk fishermen at 20 m depth off Cape May, Cape May County, New Jersey. This large left-handed whelk is harvested off New Jersey as one of the sources of the Italian molluscan seafood delicacy, *scungilli*. This large sinistral whelk ranges from southern New Jersey southward to northeastern Florida.

Figure 3.10 Northern Bay Scallop.
A, B= *Argopecten irradians irradians* (Lamarck, 1819), length 47 mm; **C, D=** *Argopecten iradians irradians* (Lamarck, 1819), length 46 mm; **E=** *Argopecten irradians irradians* (Lamarck, 1819), red color form, length 45 mm. Found in *Zostera* Eel Grass beds in quiet, protected shallow coastal lagoons, from Cape Cod to southern New Jersey. This northern subspecies represents the typical *irradians,* and it is part of a subspecies cluster that includes *irradians concentricus* (Georgian Subprovince of the Carolinian Province), *irradians taylorae* (Floridan and Suwannean Subprovinces of the Carolinian Province), and *irradians amplicostatus* (Texan Subprovince of the Carolinian Province). See Petuch and Berschauer, 2025 for details on this complex. This is the only endemic mollusk in the entire Virginian Province. Valves of the extinct *irradians sablensis* are frequently found on Sable Island off Nova Scotia.

Figure 3.11 Common Quahog Clam.
A, B= *Mercenaria mercenaria* (Linnaeus, 1758), length 95 mm; **C, D=** *Mercenaria mercenaria* (Linnaeus, 1758), 86 length mm; **E=** *Mercenaria mercenaria* (Linnaeus, 1758), length 82 mm. Found on intertidal sand flats from the Gulf of St. Lawrence, Canada to North Carolina.

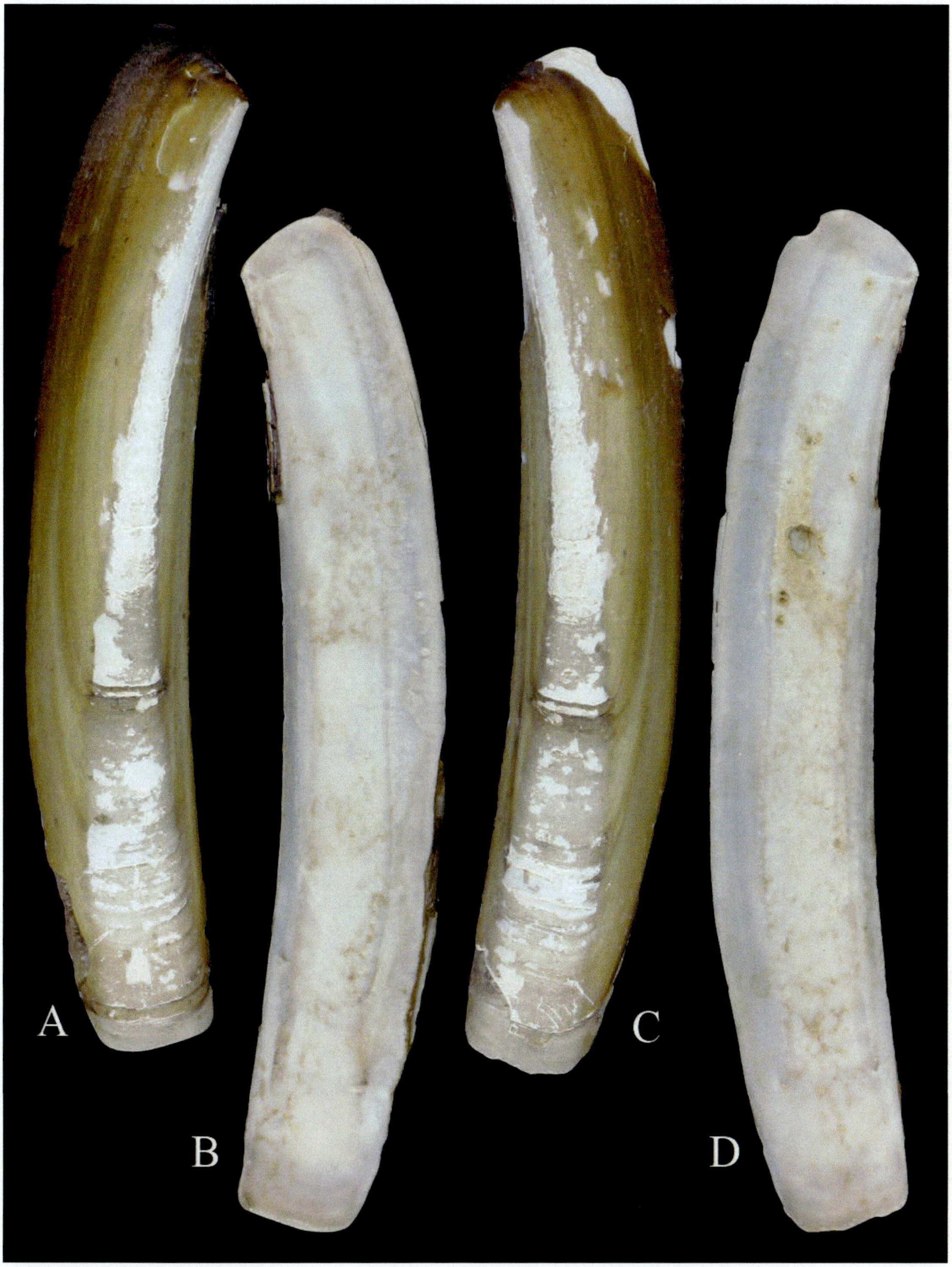

Figure 3.12 Green Razor Clam.
A, B= *Ensis directus* (Conrad, 1844), length 117 mm; **C, D=** *Ensis directus* (Conrad, 1844), length 115 mm. Found on intertidal sand flats from Labrador, Canada to North Carolina.

View of a fog bank drifting ashore onto a cobblestone beach in Cumberland County, Maine (Photo courtesy of Bryan Peña).

CHAPTER 4. The Acadian Province

Named for the Acadia Region of New France (Nouvelle-France, 17th and 18th Centuries), which today encompasses the Canadian Provinces of Nova Scotia, New Brunswick, and Prince Edward Island, the Gaspe Peninsula of Quebec, and eastern Maine as far as the Kennebec River, the Acadian Molluscan Province contains the southernmost cold water, boreal fauna found along the eastern United States. Centered on the Gulf of Maine and Bay of Fundy and the offshore Georges Banks, the Acadian Province extends northward to Labrador and the Cabot Strait and Gulf of St. Lawrence regions (Figure 4.1). The entire province is bathed in cold, subarctic water from the southward flowing **Nova Scotia Current**, a branch of the **Labrador Current**. Within the Gulf of Maine, a gyre system has become established, with the **Eastern Maine Coastal Current** flowing southward out of the Bay of Fundy where it later connects with the **Western Maine Coastal Current**. From there, the combined coastal currents flow eastward around the Wilkinson Basin out to the offshore Georges Banks, where they are deflected northward over the Georges and Browns Banks. This gyre system maintains a constant cold-water environment within the Gulf of Maine and has created a stable marine climate for the resident organisms.

Like the Virginian Province south of Cape Cod, the molluscan fauna of the Acadian Province is impoverished, especially when compared to that of the Carolinian Province. Unlike the transitional fauna of the Virginian Province, however, the Acadian Province contains some very distinctive endemic gastropods that are characteristic of this bioregion. The most noteworthy of these is seen in the stromboidean family Aporrhaidae, where three subspecies of the American Pelican's Foot Shell, *Arrhoges occidentalis*, occur in different areas of the province. Of these taxa, the typical *A. occidentalis* (Figure 4.7) has the widest range and occurs on muddy, soft-bottomed seafloors from Greenland south to Nova Scotia. In the far northwest, within the Labrador Sea, a slender, high-spired subspecies was discovered and given the name *A. occidentalis labradorensis* by Johnson in 1930 (Figure 4.8). South of Nova Scotia and throughout the Gulf of Maine, the smaller and heavily knobbed subspecies, *A. occidentalis mainensis*, is the typically encountered Pelican's Foot (Figure 4.9). South of Cape Cod, *A. occidentalis mainensis* undergoes bathymetric submergence and follows the thermocline down into the Bathyal Zone, all the way south to off South Carolina. These classically Acadian stromboideans are the "crown jewels" of the entire fauna and are much sought-after by collectors and researchers. Future studies may show that these three subspecies of the American Pelican's Foot actually represent three separate, but distinct, sibling species.

Besides the endemic aporrhaids, other gastropods that are restricted to the Acadian Province include the buccinids *Mohnia simplex, Neptunea decemcostata* (Figure 4.13), *Colus stimpsoni*, and *Colus pygmaeus*, the ptychatractid *Ptychatractus ligatus* (Figure 4.12), and the naticid *Euspira triseriata* (Figure 4.10). A distinct form or subspecies of *Neptunea decemcostata*, named *turnerae* by Clarke (Figure 4.13), is confined to the central Gulf of Maine and is most commonly collected around Mt. Desert Island. The rocky coastlines that typify the Acadian Province (shown here on Figures 4.2, 4.3, and 4.4) support an interesting fauna of hard substrate intertidal mollusks, some of which include gastropods such as the littorinids *Littorina littorea*, *L. saxatilis*, and *L. obtusata* (all three shown here on Figure 4.6), the acmaeid limpet *Collisella testudinalis* (Figure 4.5), and the muricid *Nucella lapillus* (Figure 4.11), and attached bivalves such as the mytilid

DOI: 10.1201/9781003732983-4

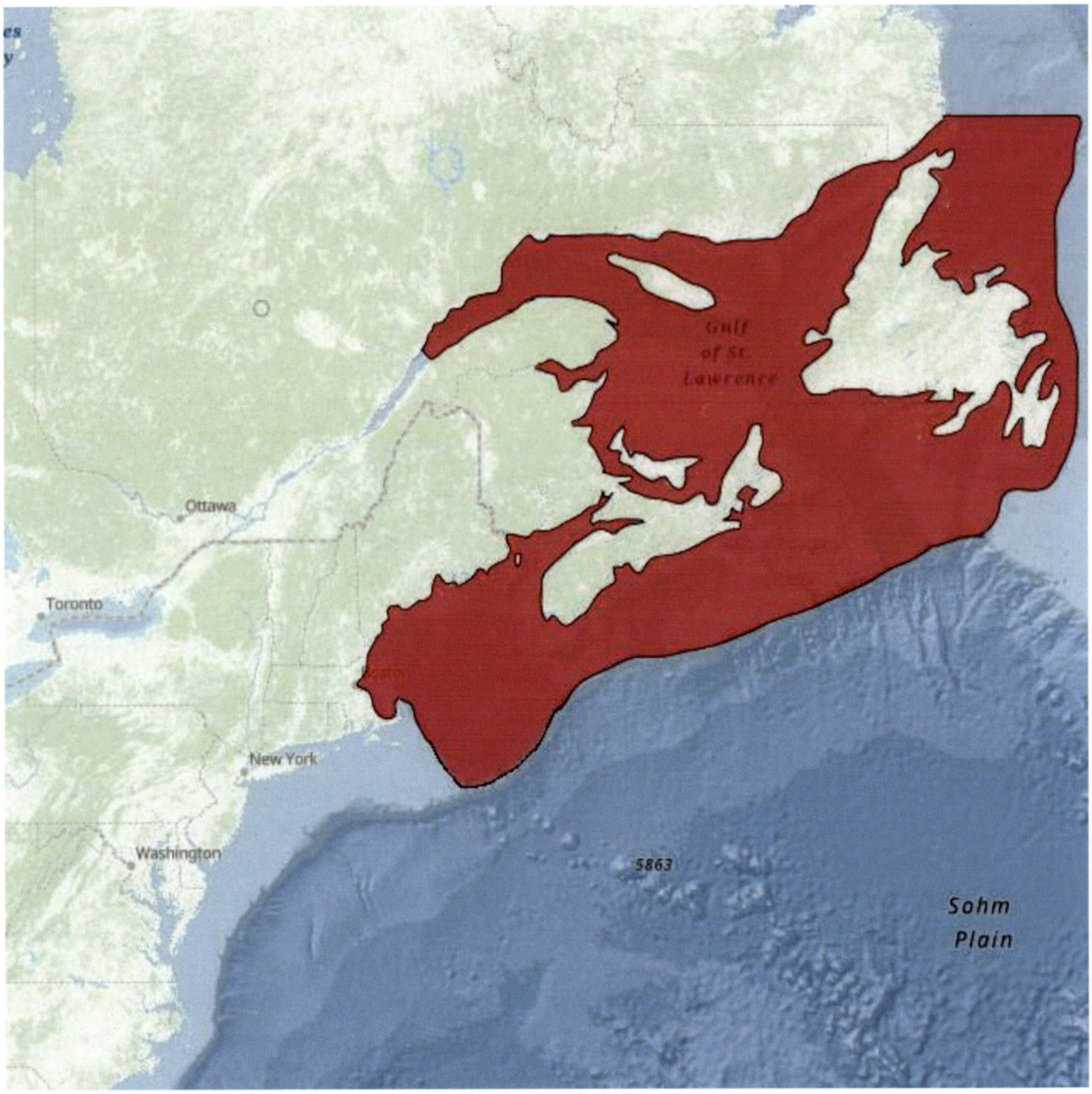

Figure 4.1 Map of the Acadian Molluscan Province.

Mytilus edulis, the modiolid *Modiolus modiolus*, and the hiatellid *Hyatella arctica*. The dense *Zostera* Eel Grass beds, that grow in shallow quiet areas between islands, once supported aggregations of the unusual sea grass limpet, *Lottia alveus* (Conrad, 1831), which lived on the individual blades. This odd, elongated limpet became extinct in the late 1920s, the result of the destruction of the Eel Grass beds along New England and the Canadian Maritime Provinces. This is the first documented extinction of a modern marine mollusk in Eastern American coastal waters.

Approximately half of the mollusks found in the Acadian Province are widespread boreal species, with many having circumpolar distributions around the world (see Bousfield, 1960). Some of these include the turritellid *Tachyrhynchus erosus* and the mytilid *Mytilus edulis*, while many others occur on both sides of the North Atlantic, such as the buccinid *Buccinum undatum*, theretemohniid *Retemohnia caelata,* and the muricid *Nucella lapillus*. During the peak glacial stages of the late Pleistocene, mollusks like these spread across the North Atlantic during times of

Figure 4.2 View of the Rocky Intertidal Zone in the Maine Coastal Islands National Wildlife Refuge (photo courtesy of Bryan Peña).

lowered sea levels and extensive ice bridges. Some, such as the muricid *Boreotrophon clathratus*, evolved in the North Pacific and Bering Sea and migrated across the Arctic Ocean, under the floating ice sheet. This same migratory pattern is seen in many of the North Atlantic echinoderms, like the multiarmed sun sea star *Crossaster papposus* and the Green Sea Urchin, *Strongylocentrotus droebachiensis*; all of which evolved in the North Pacific and migrated into the North Atlantic, via the Arctic Ocean, during the late Pleistocene. The Acadian Province molluscan fauna can now be seen to be a composite of several different malacofaunas, including endemic species that evolved autochthonously, North Pacific migrants, North Atlantic widespread taxa, and circumboreal Holarctic taxa. These are listed here in the following table:

Figure 4.3 View of the Eroded Rocky Headland of Arcadia National Park, Maine (photo courtesy of Bryan Peña).

SOME CHARACTERISTIC MOLLUSKS OF THE ACADIAN PROVINCE

Gastropoda

Acmaeidae

Collisella testudinalis (Müller, 1776)

Calliostomatidae

Calliostoma occidentale (Mighels and Adams, 1842)

Margaritidae

Margarites costalis (Gould, 1841)
Margarites groenlandicus (Gmelin, 1791)
Margarites olivaceus (Brown, 1827)
Margarites striatus (Leach, 1841)

Margarites vahlii (Moller, 1842)

Littorinidae

Littorina littorea (Linnaeus, 1758)
Littorina obtusata (Linnaeus, 1758)
Littorina saxatilis (Olivi, 1792)

Turritellidae

Tachyrhynchus erosus (Couthouy, 1838)

Aporrhaidae

Arrhoges occidentalis (Beck, 1836)
Arrhoges occidentalis labradorensis (Johnson, 1930)
Arrhoges occidentalis mainensis (Johnson, 1926)

Naticidae

Amauropsis islandica (Gmelin, 1791)
Bulbus smithii (Brown, 1839)
Euspira heros (Say, 1822)
Euspira pallida (Broderip and Sowerby, 1829)
Euspira triseriata (Say, 1826)

Muricidae

Boreotrophon clathratus (Linnaeaus, 1758)
Nucella lapillus (Linnaeus, 1758)

Buccinidae

Buccinum undatum Linnaeus, 1758
Mohnia simplex (Verrill, 1884)
Neptunea decemcostata (Say, 1826)
Neptunea decemcostata turnerae A.H. Clarke, 1956

Colidae

Colus pubescens (Verrill, 1882)
Colus pygmaeus Gould, 1841)
Colus stimpsoni (Mörch, 1867)

Retimohniidae

Retimohnia caelata (Verrill, 1884)

Ptychatractidae

Ptychatractus ligatus (Mighels and Adams, 1842)

Mangeliidae

Oenopota bicarinata (Couthouy, 1838)
Oenopota decussata (Couthouy, 1839)
Oenopota harpularia (Couthouy, 1838)
Propebela turricula (Montagu, 1803)

Bivalvia

Pectinidae

Chlamys islandica (Müller, 1776)
Placopecten magellanicus (Gmelin, 1791)

Mytilidae

Mytilus edulis (Linnaeus, 1758)

Modiolidae

Modiolus modiolus (Linnaeus, 1758)

Carditidae
Cyclocardia borealis (Conrad, 1831)
Cardiidae
Serripes groenlandicus (Bruguiere, 1789)
Arcticidae
Arctica islandica (Linnaeus, 1767)
Astartidae
Astarte castanea (Say, 1822)
Myidae
Mya arenaria Linnaeus, 1758
Hiatellidae
Hiatella arctica (Linnaeus, 1767)

Large accumulations of *Nucella lapillus* and *Littorina littorea* washed up onto the beach at Peaks Island, Maine (photo courtesy of Bryan Peña).

Figure 4.4 View of the Rocky Intertidal Zone at Petit Manan National Wildlife Refuge, Maine (photo courtesy of Bryan Peña).

Iconography of Acadian Province Species

Figure 4.5 Northern Top Shell and Limpets.
A, B= *Collisella testudinalis* (Müller, 1776), length 20 mm; **C, D**= *Collisella testudinalis* (Müller, 1776), length 18 mm; **E, F**= *Calliostoma occidentale* (Mighels and Adams, 1842), length 14 mm; **G**= *Calliostoma occidentale* (Mighels and Adams, 1842), length 12 mm. These Sub-Arctic species range from Labrador south to Long Island, New York.

Figure 4.6 Periwinkles of the Acadian Province.
A, B= *Littorina littorea* (Linnaeus, 1758), length 17 mm (ranges from Labrador to Maryland); **C=** *Littorina littorea* (Linnaeus, 1758), length 18 mm; **D, E=** *Littorina obtusata* (Linnaeus, 1758), length 11 mm; **F=** *Littorina obtusata* (Linnaeus, 1758), length 10 mm; **G, H=** *Littorina saxatilis* (Olivi, 1792), length 14 mm; **I=** *Littorina saxatilis* (Olivi, 1792), length 13 mm. In North America, *L. obtusata* and *L. saxatilis* both range from Greenland and Labrador south to Cape May, New Jersey.

Figure 4.7 American Pelican's Foot.
A, B= *Arrhoges occidentalis* (Beck, 1836), length 43 mm; **C, D=** *Arrhoges occidentalis* (Beck, 1836), length 38 mm. This Acadian Province and Boreal Province stromboidean gastropod ranges from Greenland and the Labrador Sea, the Nunavut Archipelago of Canada, south to Newfoundland, New Brunswick, Prince Edward Island, and Cape Breton Island. This is the typical subspecies of *occidentalis*.

Figure 4.8 Labrador Pelican's Foot.
A, B= *Arrhoges occidentalis labradorensis* (C. Johnson, 1930), length 52 mm; **C=** *Arrhoges occidentalis labradorensis* (C. Johnson, 1930), length 48 mm. This distinctive, high-spired subspecies of *Arrhoges occidentalis* is known only from deeper water areas off northern Newfoundland and Labrador, within the Labrador Sea of Canada. The subspecies *labradorensis* is typically pure white in color.

Figure 4.9 Maine Pelican's Foot.

A, B= *Arrhoges occidentalis mainensis* (C. Johnson, 1936), length 33 mm. This knobby southern subspecies ranges from Nova Scotia south into the Gulf of Maine and is collected off the coasts of Maine and New Hampshire. South of Cape Cod, Massachusetts, the Maine Pelican's Foot is found in very deep water, often over 300 m depths, along the continental slope (Bathyal Zone) as far south as South Carolina. Further research may show that the three subspecies of the American Pelican's Foot represent a cluster of sibling species that evolved by geographical isolation during the late Pleistocene glaciations.

Figure 4.10 Northern Moon Snails.

A, B= *Euspira heros* (Say, 1822), length 59 mm (ranges from the Gulf of St. Lawrence to deep water off North Carolina); **C=** *Euspira pallida* (Broderip and Sowerby, 1829), length 34 mm (Greenland south to deep water off North Carolina); **D, E=** *Euspira triseriata* (Say, 1826), length 28 mm (ranges from the Gulf of St. Lawrence to New York and south to off North Carolina in very deep water); **F=** *Bulbus smithii* (Brown, 1839), length 20 mm (ranges from the Gulf of St. Lawrence to Georges Bank, Massachusetts); **G=** *Amauropsis islandica* (Gmelin, 1791), length 27 mm (ranges from Greenland south to deep water off Virginia).

Figure 4.11 Northern Muricids.

A, B= *Nucella lapillus* (Linnaeus, 1758), regular color form, length 33 mm; **C**= *Nucella lapillus* (Linnaeus, 1758), purple color form, length 37 mm; **D**= *Nucella lapillus* (Linnaeus, 1758), banded color form, length 25 mm; **E**= *Nucella lapillus* (Linnaeus, 1758), banded color form, length 25 mm; **F**= *Nucella lapillus* (Linnaeus, 1758), orange color form, length 30 mm; *Nucella lapillus* ranges from Iceland to Long Island, New York; **G, H**= *Boreotrophon clathratus* (Linnaeus, 1758), length 26 mm; ranges from Iceland to the Gulf of Maine.

Figure 4.12 Whelks of the Acadian Province: Buccinidae, Retemoniidae, Ptychatractidae.
A, B= *Buccinum undatum* (Linnaeus, 1758), length 73 mm (ranges from the Arctic Ocean to New Jersey; also found in northwestern Europe); **C, D=** *Colus stimpsoni* (Morch, 1867), length 62 mm (ranges from Labrador to Long Island, New York in shallow water and south to North Carolina in deep water); **E=** *Colus pubescens* (Verrill, 1882), length 59 mm (ranges from Arctic Canada to Long Island, New York in shallow water and south to South Carolina in deep water); **F, G=** *Colus pygmaeus* (Gould, 1841), length 26 mm (ranges from the Gulf of St. Lawrence in shallow water and south to North Carolina in deep water); **H=** *Retimohnia caelata* (Verrill, 1880), length 18 mm (ranges from Iceland to Long Island, New York in shallow water and south to North Carolina; **I=** *Ptychatractus ligatus* (Mighels and Adams, 1842), length 13.5 mm, ranging from Nova Scotia to Connecticut.

Figure 4.13 Ten-Ribbed Neptune.
A, B= *Neptunea decemcostata* (Say, 1826), length 86 mm (ranges from the Grand Banks, Newfoundland, Canada to Long Island, New York in shallow water and south to North Carolina in deep water, 100-300 m depths); **C, D**= *Neptunea decemcostata* (Say, 1826), length 81 mm; **E, F**= *Neptunea decemcostata* form *turnerae* (A. Clarke, 1956), length 67 mm (a stocky and broad-shouldered variant that ranges from Grand Manan Island, New Brunswick, Canada to Mount Desert Island, Maine). This is the designated State Shell of Massachusetts.

Figure 4.14 Giant Northern Scallop.
A, B= *Placopecten magellanicus* (Gmelin, 1791), length 72 mm; **C=** *Placopecten magellanicus* Gmelin, 1791, form with radiating color pattern, length 95 mm; **D, E=** *Placopecten magellanicus* smooth form *tenuicostatus* Mighels and C.B. Adams, 1842, length 105 mm. This large scallop ranges from Labrador, Canada to Cape Cod, Massachusetts in shallow water and south to Cape Hatteras, North Carolina in deep water.

Figure 4.15 Arctic Bivalves in the Acadian Province.
A, B= *Arctica islandica* (Linnaeus, 1767), length 78 mm (ranges from Greenland to the Gulf of Maine in shallow water and south to North Carolina in deep water); **C=** *Chlamys islandica* (Müller, 1776), length 67 mm; **D=** *Chlamys islandica* (Müller, 1776), length 68 mm; **E=** *Chlamys islandica* (Müller, 1776), length 65 mm; **F=** *Chlamys islandica* (Müller, 1776), length 83 mm; ranges from Iceland to Cape Cod, Massachusetts; **G=** *Cyclocardia borealis* Conrad, 1831, length 18 mm; ranges from the Arctic Ocean to Maine in shallow water and south to Cape Hatteras, North Carolina in deep water.

CHAPTER 5. Molluscan Faunas and Catastrophic Events

The oceanographic history of the eastern coast of the United States, as preserved in the geological record of the past 35.5 million years, has revealed the presence of a continuous chain of violent and widespread catastrophic environmental changes (Petuch, 2004, 2012; Petuch and Berschauer, 2022). These rapid and devastating events are now known to have been produced by three main types of destructive influences: **1.** widespread volcanic eruptions involving flood basalts and the massive venting of gases and dust (tephra) clouds; **2.** asteroid impacts (astroblemes) and their resultant tsunamis and debris clouds; and **3.** large-scale interplanetary dust clouds and a marked attenuation of incident sunlight (insolation). Some of these terrestrial and extraterrestrial catastrophic events occurred nearly in tandem, separated from each other only by geologically small amounts of time. Each of these mechanisms initiated a cascade of environmental disasters, some of which included regional climatic cooling, glacial build-up in high latitudes, altering of oceanic currents, and lowering of sea levels. These all led to the collapse of the local marine ecosystems which, in turn, produced mass extinctions of the regional molluscan fauna. The modern Carolinian, Virginian, and Acadian Province faunas, discussed in the previous chapters of this book, all contain survivors of these catastrophic events and the most important of these are discussed in the following sections.

The Late Eocene Chesapeake Bay Asteroid Impact

The single largest catastrophic event to affect the United States eastern coast in the Cenozoic Era took place 35.5 million years ago, when a 13 kms-wide (8 miles) asteroid slammed into the shallow continental shelf off present-day Virginia, at the mouth of what is now the Chesapeake Bay (Figure 5.1). The effects of this apocalyptic event, along with those from a simultaneous impact in Siberia, drastically altered the world climate, causing mass extinctions of both terrestrial and marine organisms and marking the end of the Eocene Epoch and the Paleogene Period. By studying the distribution of the microtektites (tiny glass droplets) produced in this event, the debris field from the Chesapeake Bay Impact is now known to extend from present-day Maryland all the way to southern Argentina, with a width of over 300 km. The entire eastern United States was peppered with billions of rock fragments, both from the shattered asteroid and from the impact ejecta cloud. Numerous suspected crater features (astroblemes), produced by secondary impacts by large fragments, have been identified in North Carolina, Georgia, and northern Florida. Some of these include Lake Waccamaw and the Carolina Bays of North Carolina and the Okefenokee Swamp depression on the Florida-Georgia border. A detailed synopsis of the terminal Eocene asteroid impact and its effects on the regional geomorphology was given by Petuch and Berschauer (2022: 5-10).

The main crater of the Chesapeake Bay astrobleme was over 80 km wide and was surrounded by a bull's-eye pattern of concentric faults. The initial asteroid impact produced a series of mega-tsunamis, some as high as 500 m, that raced down the eastern coast of the United States as far

DOI: 10.1201/9781003732983-5

Figure 5.1 Map of the Chesapeake Asteroid Impact and Resultant Crater.
The outer rim of the impact structure is designated by the black line, superimposed upon a map of the present-day Chesapeake Bay area.

south as Florida and into the Gulf of Mexico. The severity of these mega-tsunamis, as they passed over the coastal lowlands and Florida Peninsula, is evident in the destruction that they produced within the local geological record. The middle and late Eocene sediments of the entire coastline were completely removed, producing a huge unconformity that is readily apparent within the stratigraphic record. In this area, younger post-impact Oligocene sediments lie directly on top of early Eocene rocks, with the middle and late Eocene being entirely absent. Because of these missing layers and the loss of their component fossils, there is little or no knowledge of the molluscan faunas that existed along the East Coast during the last half of the Eocene Epoch. This great hiatus extends from Maryland south to Florida (as the "Everglades Unconformity"; Petuch and Berschauer, 2022). During the subsequent Oligocene Epoch, sea levels were much lower than they were previously, and few marine deposits were deposited along the entire coast; little is known about the east coast molluscan faunas from that time.

Miocene Oceanography and Paleoclimates

The marine climates of Miocene North America are now known to have fluctuated wildly throughout the epoch (23 Ma to 5.3 Ma), with extensive warm periods punctuated by rapid-onset cooling events (Petuch, 1993, 1995; Farley *et al.,* 2006; Oleinik *et al.,* 2008). In North America, no place is better suited for studying these extreme climatic fluctuations than the eastern coast of the United States. Centered on Chesapeake Bay (Maryland, Virginia, and Delaware), this area contains some of the richest and best-preserved fossil beds found anywhere on the continent. The fossiliferous strata of the Chesapeake Miocene are also temporarily extensive, encompassing the entire Miocene Epoch, from the Aquitanian to the Messinian Ages. The most characteristic

Epoch	M.a.	Age	Province and climatic optima	Catastrophic event
PLIOCENE		Zanclean	Caloosahatchian Province	M.a.
MIOCENE	5.3	Messinian	Transmarian Province	5.0 Kara-Kul Impact
	7.1		LMCO ➡	
		Tortonian		8.0 Veritas Dust Cloud
			ETCO ➡	
	11.2	Serravallian		11.7 Ewing Impact 12.0 Puffin Impact 14.5 Ries-Steinheim Impacts
	14.8	Langhian	MMCO ➡	
	16.4	Burdigalian		18.0 Flood Basalts ?
	20.5	Aquitanian		
OLIGOCENE	23.8	Chattian	Prototransmarian and Antiguan Provinces	

Figure 5.2 Chronologic chart showing the timeline of Miocene asteroid impacts and interplanetary dust clouds, along with the three main Miocene climatic optima and their accompanying times of increased biodiversity. Taken from Petuch, 2012. **MMCO=** Middle Miocene Climatic Optimum (and the deposition of the Calvert Formation malacofauna); **ETCO=** Early Tortonian Climatic Optimum (and the deposition of the St. Mary's Formation malacofauna); **LMCO=** Late Miocene Climatic Optimum (and the deposition of the Eastover Formation malacofauna).

genera and species of mollusks, both gastropods and bivalves, are shown in the Iconography section at the end of Chapter 6 of this book.

Biogeographically, the eastern North American molluscan faunas were situated within an appropriate range of latitudes to give maximum insight into the ecological effects of Miocene temperature fluctuations. Although normally occurring in temperate-water conditions and containing temperate endemic mollusks, the Chesapeake molluscan faunas became inundated with tropical-subtropical migrants during warm times and, conversely, with boreal taxa during cold times (Petuch, 1988, 2004, 2012; Petuch and Drolshagen, 2009). These faunal shifts were accompanied by dramatic extinctions and extirpations, reflecting the rapidly changing marine climates. The Miocene molluscan faunas illustrated in this book are from four different formations within the Chesapeake Group and one separate formation from North Carolina. These include: the Silverdale Member of the **Belgrade Formation** (Aquitanian Age); the **Calvert Formation** (with its Fairhaven, Plum Point, and Calvert Beach Members, Burdigalian and Langhian Ages); the **Choptank Formation** (with its St. Leonard, Drumcliff, and Boston Cliffs Members, Serravallian Age); the **St. Mary's Formation** (with its Conoy, Little Cove Point, Windmill Point, and Chancellor's Point Sandstone Members, late Serravallian and Tortonian Ages); and the **Eastover Formation** (with its Claremont Manor and Cobham Bay Members, late Tortonian and Messinian Ages).

Based on faunal composition, three long warm intervals can be seen to have occurred in eastern North America during the Miocene: one extending from the late Burdigalian to the late Langhian (approximately 16-18 Ma) and corresponding to the **Middle Miocene Climatic Optimum** (MMCO) of Oleinik *et al.,* 2008; one extending from the latest Serravallian to the early Tortonian (approximately 11.4-10 Ma); and one extending throughout the entire Messinian (approximately 7-5 Ma). These warm intervals correspond to the deposition of the Calvert, St. Mary's, and Eastover Formations, respectively. Bracketing these three major warm intervals were four cold intervals: one extending from the Aquitanian to late Burdigalian (approximately 23-16 Ma); one during the Serravallian (approximately 14.4-11.5 Ma); and one during the late Tortonian (approximately 8 Ma). In Maryland and Virginia, these correspond to the depositions of the Old Church (poorly preserved and rarely exposed) and Choptank Formations, the unconformity between the St. Mary's and Eastover Formations, and the unconformity between the Eastover and Yorktown Formations (Miocene-Pliocene Boundary). These Miocene warm climatic optima and their corresponding stratigraphic units are shown here on Figure 5.2.

Geochronological studies of astroblemes and cosmological events from around the world have shown that several major asteroid impacts took place during the Miocene (Shoemaker and Chao, 1961; Gorter and Bayford, 2000; Glickson, 2005; Farley *et al.,* 2006). Giant gas and dust clouds passing through our solar system also had climatological repercussions, as they decreased the amount of incident sunlight (insolation) reaching the Earth. The most important and best documented of these Miocene extraterrestrial catastrophic events are discussed in the following sections, and the mollusks that were affected by these events are illustrated in the Iconography at the end of Chapter 6.

Figure 5.3 View of a highly fossiliferous exposure of the Choptank Formation, in a cliff near Matoaka Cottages, St. Leonard, Calvert County, Maryland, Chesapeake Bay. The lower bed is the St. Leonard Member (Shattuck Zone 17), and the overlying unfossiliferous bed is the Drumcliff Member (Shattuck Zone 18). At most places along the western shore of Chesapeake Bay, Shattuck Zone 18 contains a sparse fauna of fossil mollusks. This location, however, is devoid of any fossils and is composed of a thick clay unit. The sharp contact between the two members implies that there was an abrupt drop in sea level during the late Serravallian Age, and a subsequent period when the exposed seafloor (now Zone 17) was eroded under subaerial conditions. This resulted in the removal of the seafloor material and the production of an erosional unconformity. This was most probably caused by the worldwide climatic cooling due to the combined Puffin and Ewing asteroid impacts. A large soil slump, caused by the current rising sea level, can be seen farther down the beach.

The Mid-Langhian Radcliffe Wave

Named for the Radcliffe Institute for Advanced Study at Harvard University, where it was first recognized, the Radcliffe Wave is a giant cohesive arm of interstellar gas stretching for over 9,000 light years across the Milky Way galaxy. Even though the closest section of this immense stream of gas and star-forming regions is now 400 light years away, research has shown that the Radcliffe Wave passed directly over, and through, our solar system during the middle of the Langhian Age, immediately after the warm time of the Middle Miocene Climatic Optimum. Prior to the passage of the Radcliffe Wave, the eastern coast of the United States had a subtropical marine climate, as evidenced by the presence of classic tropical-warm water index genera such as *Xenophora* (Xenophoridae), *Americoliva* (Olividae), *Ficus* (Ficidae), and *Amaea* and *Cirsotrema*, (both Epitoniidae) (examples shown in Chapter 6 of this book). After the wave had passed at the end of Langhian time, several important endemic muricid groups had also become extinct (see Petuch and Berschauer, 2025), including the ecphora ocinebrine subgenera *Eccentrella*, *Matoaka*, *Recurvecphora*, and the genus *Siphoecphora*. Their extinction during the deposition of the Calvert Beach Member of the Calvert Formation demonstrates that many coastal ecosystems collapsed due to cold-water temperatures.

The Serravallian Ries-Steinheim, Puffin, and Ewing Impacts

By the beginning of Choptank Formation deposition in the Serravallian Age, warm marine climates had returned to the coastal areas of the eastern United States. The **Radcliffe Wave** had moved on away from the solar system and incident sunlight returned to its prior level of insolation. The warming of the marine climate is evidenced by the appearance of a rich new molluscan fauna in the basal beds of the Choptank Formation, which contained subtropical elements such as the cancellariid genus *Ventrilia* and the scaphelline volutid genus *Volutifusus*. By the earliest Serravallian Age, however, the climate quickly degenerated, and a prolonged time of global cooling and wildly fluctuating sea levels began (Petuch, 2004; Petuch and Drolshagen, 2009). In total, the destructive effects of three separate cooling episodes, in quick succession, can be seen in the Serravallian fossil record of Maryland. The first of these, at around 14.5 Ma, corresponds directly to the **Ries-Steinheim Double Impact** in Bavaria (Shoemaker and Chao, 1961). The albedo effect produced by this double asteroid impact was apparently strong enough to plunge the entire Northern Hemisphere into a long-lived severe cold time that resulted in the extirpation (regional extinction) of all tropical taxa in high latitudes.

Because of the lingering effects of the Ries-Steinheim Impact, the Serravallian molluscan fauna of Maryland was much less species-rich than the faunas of the older Langhian Age. Out of the 65 macrogastropod genera found in the Calvert Formation (Langhian Age), 19 (29%), disappeared from the Maryland fossil record at this time (Petuch, 2012: 974). Only 25 of these genera survived into the time of Choptank deposition and these were dominated by cold-tolerant taxa such as the naticids *Euspira* and *Euspirella*, the turritellid *Mariacolpus*, the ocinebrine muricids *Chesathais* and *Trisecphora*, and the busyconid *Sycopsis* (all shown in Chapter 6 of this book). The boreal genus *Euspira* was abundant during this time and dominated the Choptank faunas, attesting to the colder marine climate during the early Serravallian. This mid-Miocene cooling was further exacerbated by a second asteroid impact during Serravallian time around 12 Ma; the **Puffin Impact** off northwestern Australia (Gorter and Bayford, 2000). Being a continental shelf impact, a large amount of ejecta would have been thrown into the stratosphere, being carried around the world and again increasing the Earth's albedo.

A third asteroid impact occurred soon after the Puffin Impact, at 11.7 Ma, on the Pacific seafloor near the equator southeast of Hawaii. Referred to as the **Ewing Impact**, this was the largest of the three mid-Miocene impacts and produced the most widespread effects. As pointed out by Abbott *et al.* (2003) and Leung *et al.* (2004), silicon carbide spherules traced to the impact were found as far away as Arizona. This indicates that a tremendous amount of dust and debris was blasted into the stratosphere or higher, increasing the Earth's albedo and lowering worldwide temperatures. The data gleaned from such a massive impact readily explains the abrupt climatic degeneration and large extinction at the end of the Choptank Formation deposition. The abrupt extinction of the muricid genera *Ecphorosycon*, *Trisecphora*, *Chesathais*, *Ecphora (Rhombecphora)*, *Trisecphora (Matoaka)*, and *Patuxentrophon*, and their disappearance from the East Coast fossil record, demonstrates the extremely strong global cooling that was produced by an impact on the other side of the planet.

The Mid-Tortonian Veritas Interplanetary Dust Cloud

At the end of the middle of the Tortonian Age, 8.2 Ma, the newly recovered marine ecosystems of eastern North America underwent yet another collapse, resulting in the extinction of almost one-third of the macrogastropod fauna (26.6%, with 16 genera becoming extinct out of a total of 60). This abrupt truncation of the molluscan fauna is obvious when the species-richness of the early Tortonian St. Mary's Formation (Windmill Point Member) fauna is compared with that of the subsequent Eastover Formation (Claremont Manor Member) (Petuch and Drolshagen, 2009; discussed in Chapter 6 of this book). Particularly hard-hit were the families Fasciolariidae (with the extinction of *Mariafusus*, *Conradconfusus* and *Pseudaptyxis*), Busyconidae (with the extinction of *Sycopsis, Sycotypus*, and *Turrifulgur*), and the superfamily Turroidea (with the extinction of *Chesaclava*, *Chesasyrinx*, *Hemipleurotoma*, *Mariaturricula*, *Nodisurculina*, *Sediliopsis*, and *Mariadrillia*), all of which are illustrated in Chapter 6.

This major extinction and extirpation event, which destroyed the biogeographical structure of eastern North America, correlates directly to a significant drop in incident sunlight (insolation) that was associated with an obscuring interplanetary dust cloud. Based on meteoric dust samples, Farley et al. (2006) demonstrated that two or more of the **Veritas Family** of asteroids (in the main asteroid belt) collided and disintegrated into an immense cloud of dust at 8.2 Ma. This thick dust cloud settled into an Earth-crossing orbit, obscuring the sun and greatly diminishing the incident solar radiation for the entire planet. The lowered insolation resulted in severe and rapid cooling events around the world and the catastrophic collapse of many shallow-water marine ecosystems. The subtropical oceanic climate of the eastern United States was plunged, almost instantaneously, into a cold time that persisted for hundreds of thousands of years. The complete collapse of the rich early Tortonian St. Mary's fauna, with the abrupt disappearance of tropical groups like *Semicassis*, *Eudolium*, *Laevityphis*, *Hesperisternia*, *Mitra*, *Ventrilia*, and *Gradiconus* points to a rapid-onset cooling event. The quickness of this cooling event did not allow enough time for these temperature-sensitive groups to migrate southward, and they were trapped in a marine climate that was cooling faster than they could adapt, leading to their extinction and extirpation.

The Late Messinian Kara-Kul Impact

At the very end of the Miocene, one of the most significant marine faunal extinctions of the entire Neogene occurred worldwide during the latest Messinian Age (Petuch, 1993). At this time, the Northern Hemisphere quickly cooled and ice sheets began to expand, resulting in a major glacioeustatic low. This sea-level drop was so severe that the Mediterranean Sea basin became cut off from the Atlantic Ocean and completely dried up (the "**Messinian Event**"). The extreme sea level low that caused the Messinian Event is represented in Maryland and Virginia by a large unconformity between the late Miocene Eastover Formation and the Pliocene Yorktown Formation (Petuch and Drolshagen, 2009: 22-23). Worldwide oceanic current patterns were also disrupted at this time, as exemplified by the changes in the Pacific North Equatorial Countercurrent that were detected in the latest Miocene beds of Panama (Collins *et al.*, 1996). This extreme eustatic low, coupled with severely lowered water temperatures, quickly annihilated the newly rebounded molluscan faunas of Maryland and Virginia, resulting in the **Transmarian Extinction** (Petuch, 2004: 261-262; discussed in Chapter 6). At this time, classic Transmarian Province genera such as the turritellid *Mariacolpus*, the naticid *Euspirella*, the busyconid *Coronafulgur*, the nassariid *Bulliopsis*, the conid *Gradiconus*, and the terebrid *Laevihastula* became extinct and the paleoprovince, as a biogeographical entity, ceased to exist.

The late Messinian Transmarian Extinction, at 5 ma, correlates exactly with the formation of the large **Kara-Kul Impact** Structure in the Pamir Mountains of Tajikistan (Gurov *et al.*, 1993; Gurov and Gurova, 1993; now a double lake, first discovered on satellite imagery). Being a central continental impact, Kara-Kul would have produced immense amounts of ejecta and aerosols which, in turn, would have greatly increased the Earth's albedo. The resultant lowered temperatures were severe enough to have cooled the northwestern Atlantic to nearly boreal conditions and to have dropped sea levels by at least 100 m. This would have been sufficient to decimate the stenothermal shallow-water Transmarian ecosystems. Chronostratigraphically, the effects of the Kara-Kul Impact were sufficient to have caused the extinction of many marine faunas around the world and the alteration of sedimentary processes. This chronologic marker is so impressive and widespread that it was recognized by most early workers, who designated it to mark the Miocene-Pliocene boundary.

Pliocene Oceanography and Paleoclimates

The Pliocene Epoch (5.3 to 2.6 Ma) is divided into two separate ages: a lower Zanclean Age that extends from 5.3 to 3.6 Ma) and an upper Piacenzian Age that extends from 3.6 Ma to 2.6 Ma. During the Zanclean Age, the eastern United States was still recovering from the environmental effects of the Kara-Kul Impact and did not return to the tropical conditions that were seen during the early Tortonian Miocene until around 3 Ma. The cooler-water conditions of Zanclean eastern North America are underscored by the presence of several cool-water molluscan genera that were prevalent organisms during the late Miocene. Some of these survivors include gastropod genera such as the muricids *Lirosoma*, *Scalaspira*, *Chesatrophon*, *Ecphora*, *Ecphora (Powhatan)*, and *Planecphora* and the scaphelline volute *Scaphellopsis* and bivalves such as *Chesapecten*. These are all dominant organisms in the lower members (Mansfield Zone 1) of the Yorktown Formation of Virginia and North Carolina (all shown in Chapter 7 of this book). These all became extinct around 3 Ma, during the late Zanclean warming time.

At the beginning of the Piacenzian Age, carbon dioxide levels began to increase in the Earth's atmosphere and the marine and terrestrial climates began to warm. The increased volcanism during this time, especially in places like Iceland, Indonesia, the western United States mountains, and the Andes, collectively led to an accumulation of atmospheric "greenhouse gasses" that raised worldwide temperatures. This warming pattern is especially evident in southern Florida, where an immense atoll-like coral reef feature, the **Everglades Pseudoatoll**, grew in the high tropical marine climate of the late Piacenzian (Petuch, 2004; Petuch and Roberts, 2007; Petuch and Berschauer, 2022). The coralline environments of the Everglades Pseudoatoll supported one of the richest molluscan faunas ever discovered anywhere else on Earth, equivalent to the living faunas of modern Philippines and Indonesia. Many of these Piacenzian tropical genera, such as *Akleistostoma* (Cypraeidae), *Pleioptygma* (Pleioptygmatidae), *Ximeniconus* and *Contraconus* (Conidae), and *Cinguloterebra* (Terebridae), migrated northward out of the pseudoatoll and populated the eastern coast of the United States with entire radiations of warm-water taxa. Some of these are illustrated and discussed in Chapter 7 of this book.

The Late Pliocene Eltanin Impact

The high tropical environments and ecosystems of the late Piacenzian were decimated by the effects of another major asteroid impact, this time in the Bellinghausen Sea, in the eastern part of the South Pacific Ocean, southwest of the southern tip of Chile. First discovered in 1981, the submerged impact site was named the **Eltanin Impact Area,** after the research vessel that first collected samples and sedimentary evidence. Because it took place in 4-5 km-deep water, no impact structure has ever been discovered, but it has been estimated that over 150 cubic kms (36 cubic miles) of seawater were evaporated during the initial strike. Multiple megatsunamis were also generated by the impact, with most being over 500 m in height. Based on sedimentary structures and iridium isotopes, it is now known that the Eltanin Impact took place 2.5 Ma, at exactly the Pliocene-Pleistocene Boundary. Of the late Neogene asteroid impacts, the Eltanin caused the greatest climatological and environmental degradation and, most importantly, ushered in the first "Ice Age" of the Pleistocene Epoch. A few hardy survivors from the older warmer water faunas managed to persist into the early Pleistocene (Gelasian Age) along the East Coast and included the muricids *Trossulasalpinx*, *Mariasalpinx*, and *Pterorhytis* and the conid *Ximeniconus*. By the end of the Gelasian Age, however, these hardy survivors became extinct and the remaining molluscan faunas were greatly impoverished. For an overview of the late Pleistocene molluscan fauna of the eastern United States, see Chapter 8 of this book.

A foggy and rainy day along the Calvert Cliffs, Calvert County, Maryland, on the western side of Chesapeake Bay. Here, near Matoaka Cottages at St. Leonard, the highly fossiliferous beds of the Choptank Formation (tan-brown shell beds) can be seen to stop abruptly and are capped with an unfossiliferous black layer. This sharp depositional break marks the worldwide climatic cooling event that was caused by the Ries-Steinheim Double Asteroid Impact of the mid-Serravallian Miocene.

CHAPTER 6. Miocene Molluscan Faunas of the Eastern United States

As shown in Chapter 5, the Miocene Epoch (23 Ma to 5.3 Ma) was a dynamic time of tremendous environmental changes taking place all over the planet. During this epoch, the Himalaya Mountains reached their peak development, and this orogeny altered weather patterns across the world; deflecting atmospheric jet streams, changing high-altitude wind patterns, and altering oceanic currents (Petuch, 1988; 2004). Extensive grasslands developed in areas that had previously been rainforests, and this led to the evolution of entirely new terrestrial mammal faunas. In the Atlantic Ocean, the closed, landlocked, northern boundary, which previously prevented cold arctic water from flowing into the North Atlantic, was breached as Greenland moved westward and a connection between the Labrador Sea and Arctic Ocean was established. This tectonic event was marked by the rapid appearance of North Pacific and Arctic invertebrates in the fossil record, particularly mollusks and echinoderms. The kelp forests (particularly *Laminaria*) that now occur in the North Atlantic can trace their ancestry back to the Arctic Ocean and North Pacific. The cooler and nutrient-rich waters of the early Miocene also allowed for the evolution of new cetacean and elasmobranch faunas, setting the stage for modern whales and sharks.

The Miocene is now known to have been the time when the first modern molluscan fauna became established along the eastern United States. Many of the living taxa shown here in Chapters 2, 3, and 4 evolved in this new oceanic system during the early Miocene. As discussed in Chapter 5, the Miocene was also a time of intermittent environmental degradation, manifested as episodes of lowered sea level and long periods of cold air and water temperatures. The most severe of these can be correlated with asteroid impacts all around the planet, demonstrating that extraterrestrial events caused many of the major extinctions of mollusks; these are also preserved in the fossil record and stratigraphic columns.

The Salisbury and Albemarle Seas

The coastal geography and geomorphology of the eastern United States began to change during the early part of the Oligocene Epoch, when tropical climatic conditions prevailed and heavy monsoonal rainfall increased the erosion of the Appalachian Mountains. This led to the formation of a wide apron of sediments along the eastern side of the mountains. Coastal lowlands such as these are vulnerable to sea-level fluctuations and oceanic erosion, and by the beginning of the Miocene (Aquitanian Age), this wide, flat coastal plain had eroded to form two large, shallow baylike epeiric seas: the **Albemarle Sea** in the south (Figure 6.1) and the **Salisbury Sea** in the north (Figure 6.2), with both separated by the intervening Norfolk Arch highlands area. Referred to by many geologists as "embayments", these two large bodies of water were actually true seas in the oceanographic sense and satisfied the four conditions for paleosea status (as defined by Petuch, 2004): **1.** they were bodies of salt water; **2.** they were bounded on at least three sides by geological structures (such as mountain ranges, arches, topographic highs, etc.); **3.** they had their own individual water masses and currents; and **4.** they contained their own distinctive marine faunas and floras, with large numbers of endemic species. Depending on sea level fluctuations, the sedimentation within the paleosea basins was frequently truncated, producing a noticeable break (unconformity) between each sedimentary depositional unit. These unconformities represent times when the basin dried out and terrestrial conditions prevailed for long periods. The

DOI: 10.1201/9781003732983-6

resultant package of sequestered sediments is referred to as a **subsea**. The Albemarle Sea and the Salisbury Sea both contain several distinct subseas, most of which contain similar molluscan faunas, with many genera and species chronologically restricted. These taxa are useful index fossils for geological correlations, and many are shown here in the Iconography at the end of this chapter.

During the first half of the Miocene, the Albemarle Sea (named for the Albemarle Sound of North Carolina) was the smaller of the two regional bodies of sea water, covering only the area of present-day eastern North Carolina (Figure 6.1). The larger and more structurally complex Salisbury Sea (named for the city of Salisbury, Maryland) to the north covered parts of four states: extending from the Great Falls of the Potomac River in Maryland and the Bluffs of Richmond, Virginia, south to the James River at Norfolk, Virginia, north to near Millville in southern New Jersey, and encompassing the entire State of Delaware and the Delmarva Peninsula (Figure 6.2). The basin of the Salisbury Sea was roughly circular and conformed to the shape of the astrobleme that was produced by the Chesapeake Bay Asteroid Impact during the latest Eocene (described in Chapter 5 of this book; see Petuch and Berschauer, 2022). During the Serravallian Age of the middle Miocene, a large delta system (the **Millville Delta**; see Petuch, 1988: 102) developed in the northern part of the Salisbury Sea and these deltaic sediments infilled roughly half of the Salisbury marine basin. By the end of the middle Serravallian Age, a combination of lowered sea levels and increased development of the Millville Delta caused massive sedimentary infilling of the Salisbury Sea, reducing its size to only half of what it was during the Langhian Age. This reduction in size, accompanied by the colder marine climate that resulted from the Puffin Asteroid Impact, led to a mass extinction of the Salisbury molluscan fauna. This is most readily seen by comparing the thriving ecphora fauna of the early and middle Serravallian Choptank Formation (Shattuck Zones 18 and 19), containing five species in five different genera and subgenera, with the late Serravallian Conoy Member of the St. Mary's Formation (Shattuck Zone 20), with only a single genus and species. For the rest of the Miocene, only a single species of ecphora occurred in any given geological formation or member (Petuch and Berschauer, 2023).

By the beginning of the Tortonian Age, the marine climate warmed considerably and a subtropical molluscan fauna was now present within the Salisbury Sea. As seen by the fauna of the St. Mary's Formation, with examples shown in the Iconography at the end of this chapter, a number of classic tropical gastropod groups were present during the early Tortonian, including *Gradiconus* (Conidae), *Eudolium* (Tonnidae), *Hesperisternia* (Pisaniidae), *Scaphellopsis* and *Volutifusus* (Volutidae), and *Cymatosyrinx* (Drilliidae). By this time, the Millville Delta had grown to such a large size that it infilled fully three-quarters of the Salisbury Sea basin, resulting in only a small crescent of marine environments existing in the southernmost section (what is now modern Virginia between the Potomac and Rappahannock Rivers; Petuch, 1988:110). The late Tortonian Albemarle Sea was nearly obsolete, represented only by a shallow embayment that extended inland for a short distance and entirely infilled with sediments from the eroding Norfolk Arch. By the beginning of the Messinian Age, the highlands of the Norfolk Arch had eroded away, and this allowed the much-reduced Albemarle and Salisbury Seas to expand and fuse into a single new marine world, the **Rappahannock Subsea** of the Albemarle Sea (see Petuch, 1988: 124). This is a classic example of "basin capture" and the new enlarged subsea housed a diverse subtropical marine fauna that contained classic warm-water groups such as the cone shell *Gradiconus* and the olive shell *Americoliva*, along with extensive bioherms of the coral *Septastrea*. These tropical faunal elements are all preserved in the Messinian Miocene Eastover

Formation of Virginia and northern North Carolina, and many examples are shown here in the Iconography at the end of this chapter. The Kara-Kul Asteroid Impact at the end of the Messinian Age (Messinian-Zanclean Boundary) caused sea levels to drop dramatically worldwide, and much of the Rappahannock Subsea dried up, resulting in the extinction of many of the characteristic Eastover taxa, including the turritellid *Mariacolpus*, the busyconid *Coronafulgur*, and the nassariid *Bulliopsis*.

Figure 6.1 Map of the Miocene Albemarle Sea during its maximum development in the Langhian Age. The northern edge of the Albemarle Sea was bound by the Norfolk Penninsula which separated it from the Salisbury Sea.

The subseas of the Miocene Albemarle and Salisbury Seas include:

ALBEMARLE SEA

Silverdale Subsea: Aquitanian Miocene (Containing the fauna of the Haywood Landing Member of the Belgrade Formation).

Pamlico Subsea: Latest Burdigalian and Langhian Miocene (Containing the fauna of the Pungo River Formation).

Rappahannock Subsea: Late Tortonian and Messinian Miocene (Containing the fauna of the Eastover Formation).

SALISBURY SEA
Old Church Subsea: Aquitanian Miocene (Containing the fauna of the Old Church Formation).
Calvert Subsea: Latest Burdigalian and Langhian Miocene (Containing the fauna of the Calvert Formation).
Patuxent Subsea: Early and Middle Serravallian Miocene (Containing the fauna of the Choptank Formation).
St. Mary's Subsea: Late Serravallian and early Tortonian Miocene (Containing the fauna of the St. Mary's Formation).

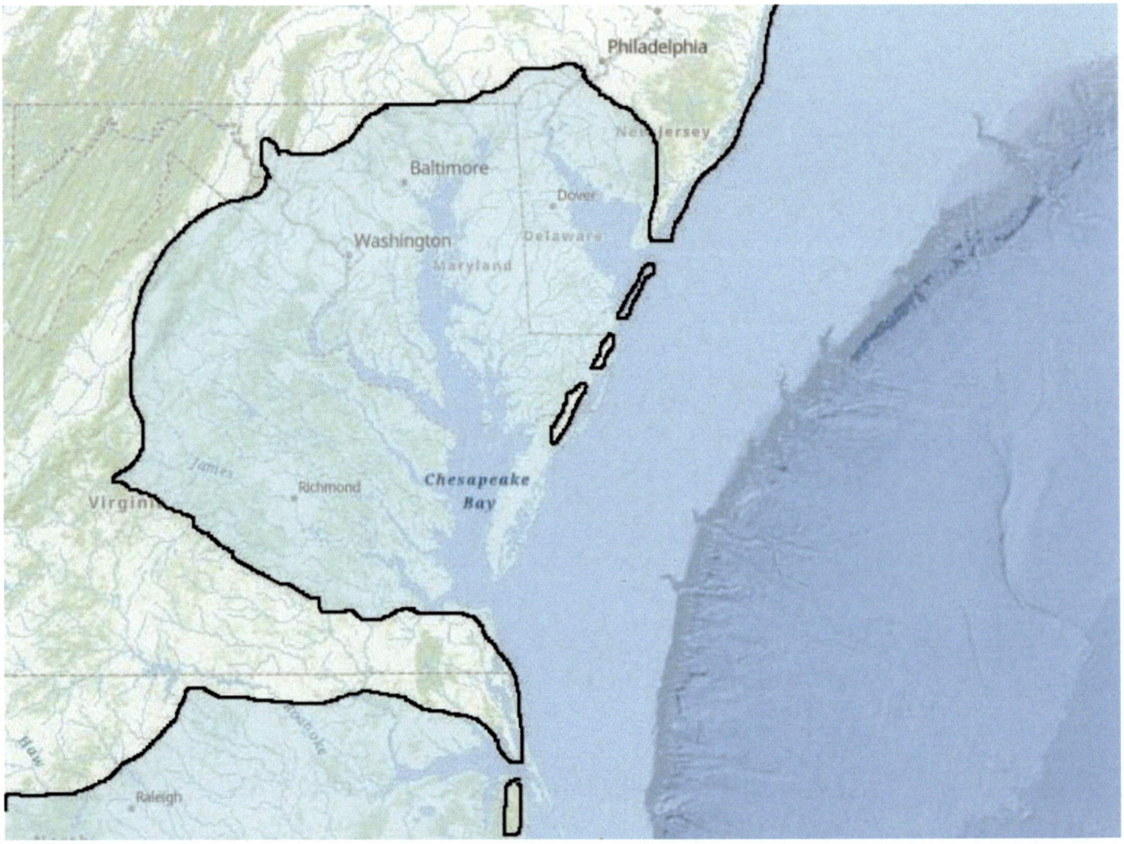

Figure 6.2 Map of the Miocene Salisbury Sea during its maximum development in the Langhian Age. The southern part of the basin was structurally bound by the Norfolk Peninsula and the shape of the sea conforms roughly to the shape of the asteroid impact crater. The northeast corner of the Salisbury Sea contained the Millville Delta whose sediments were producing the Delmarva Peninsula.

The Transmarian Province

Named for the St. Mary's River of St. Mary's County, Maryland, this Miocene molluscan faunal province is the ancestor of the modern Carolinian Province that is described in Chapter 1 of this book. Like the Carolinian Province and its Georgian Subprovince, the Transmarian Province is defined by percentages of endemism seen in the component gastropod index families and subfamilies, as defined by Petuch, 2013 and Petuch and Berschauer, 2022. Considering that an endemism level of 50% is the minimum for full provincial status, the Transmarian Province is extremely robust when compared to contemporary provinces to the south, in Florida and the Caribbean, with an Index of Provinciality of around 90% at the generic level. This extremely high level of endemism underscores the different, and long-lived, ecological and oceanographic conditions that existed along the east coast at that time. During its maximum development in the Langhian, the Transmarian Province extended from New Jersey to southern North Carolina and encompassed the Albemarle and Salisbury Seas and their adjacent coastal lagoons.

Based upon the ranges of endemic species and species complexes within the provincial index genera, the Transmarian Province can be further subdivided into two distinct subprovinces, including the southern **Pungoian Subprovince** (named for the Pungo River Formation), which was confined to the Albemarle Sea, and the northern **Calvertian Subprovince** (named for the Calvert Formation), which was primarily in the Salisbury Sea. Although over 30% of the species were different between the two subprovinces, both biogeographical regions shared the same characteristic index genera. These are listed here and are illustrated in the Iconography at the end of this chapter.

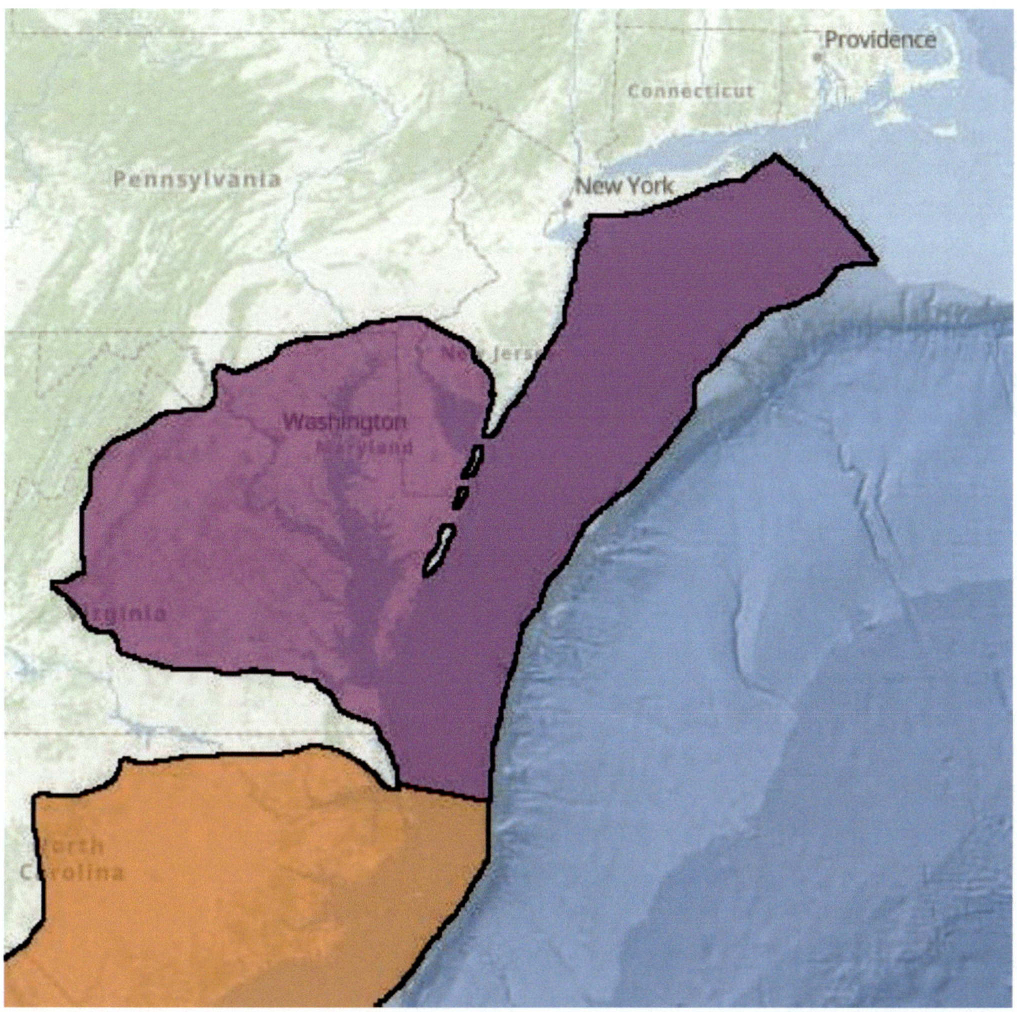

Figure 6.3 Map of the Transmarian Province and its Subprovinces during the Miocene. Gold color denotes the Pungoian Subprovince (named for the Pungo River Formation); lavender color denotes the Calvertian Subprovince (named for the Calvert Formation).

INDEX GENERA FOR THE TRANSMARIAN PROVINCE

Gastropoda

Turritellidae

Calvertitella Petuch, 1988
Mariacolpus Petuch, 1988

Naticidae

Euspirella Petuch and Drolshagen, 2010
Poliniciella Petuch, 1988

Cassidae

Mariacassis Petuch, 1988

Muricidae

Chesathais Petuch, 1988
Chesatrophon Petuch, 1988
Ecphora (Rhombecphora) Petuch and Berschauer, 2023
Ecphorosycon Petuch, 1988
Lirosoma Conrad, 1862
Mariasalapinx Petuch, 1988
Patuxentrophon Petuch, 1992
Scalaspira Conrad, 1862
Siphoecphora Petuch, 1988
Stephanosalpinx Petuch, 1988
Trisecphora Petuch, 1988
Trisecphora (Eccentrella) Petuch and Berschauer, 2023
Trisecphora (Matoaka) Petuch and Berschauer, 2023
Trisecphora (Recurvecphora) Petuch and Berschauer, 2023

Fasciolariidae

Conradconfusus Snyder, 2002
Mariafusus Petuch, 1988
Pseudaptyxis Petuch, 1988

Busyconidae

Coronafulgur Petuch, 2004
Sycofulgur Marks, 1950
Sycopsis Conrad, 1867
Turrifulgur Petuch, 1988

Nassaridae

Bulliopsis Conrad, 1862

Cancellariidae

Cancellariella Martin, 1904
Marianarona Petuch, 1982
Mariasveltia Petuch, 1982

Conidae

Gradiconus da Motta, 1991

Turridae

Chesasyrinx Petuch, 1988
Hemipleurotoma Cossmann, 1899
Mariaturricula Petuch, 1988
Nodisurculina Petuch, 1988

Mangeliidae
Calverturris Petuch, 1993
Mariadrillia Petuch, 1988
Transmariaturris Petuch, 1993
Drilliidae
Chesaclava Petuch, 1988
Sediliopsis Petuch, 1988

Molluscan Faunas of the Miocene Albemarle and Salisbury Seas

Within the Transmarian Province, several groups of prominent mollusks underwent large species radiations throughout the Miocene. These spurts of evolution took place during the three climatic optima, which occurred between the extinctions caused by asteroid impacts (as outlined in Chapter 5). The singlemost spectacular evolutionary explosion that took place within the molluscan fauna of the Transmarian Province was seen in the Tribe Ecphorini (the Ecphoras), a large group of endemic ocenebrine muricids (described and illustrated by Petuch and Berschauer, 2023). These large and ribbed murex shells dominated the local ecosystems and were the primary predators of the diverse and abundant Transmarian bivalve fauna; 39 species of Miocene ecphoras are also shown here in the Iconography at the end of this chapter. The Tribe Ecphorini reached its evolutionary peak during the Langhian and early Serravallian Ages, as every geologic member of the Calvert and Choptank Formations contain 5 or 6 different species in at least 4 different genera. The richest single fauna of ecphoras is seen in the Plum Point Member (Shattuck Zone 10) of the Calvert Formation, Langhian Age, where 8 different species in 5 genera and 3 subgenera occured together in a single unit. After the Ewing and Puffin Impacts of the middle and late Serravallian Ages, the ecphoras underwent a major extinction event, with every subsequent Miocene stratigraphic unit containing only a single species of the genus *Ecphora*.

The abundance of small bivalve prey seen in the Chesapeake Miocene formations supported a large fauna of molluscivorous gastropods, particularly the Moon Snails (family Naticidae; see Petuch, 1988; Petuch and Drolshagen, 2009). These naticids were among the dominant predators within the local ecosystems, as evidenced by the abundant drilled holes in bivalve and turritellid shells. Judging from the size and shape of the predation holes, the Chesapeake Miocene naticid fauna exhibited niche partitioning; with large species, such as those in the genera *Euspira* and *Neverita (Glossaulax)*, feeding on large thick-shelled bivalves such as venerids, crassatellids, and and pectinids, and with small species, such as those in the genera *Euspirella*, *Poliniciella*, and *Sinum*, feeding on small mactrids and tellinids. The endemic Transmarian turritellid genus *Mariacolpus* formed extensive beds, composed of millions of individuals, that covered the seafloor at many localities. These small turritellids also served as prey items for the larger naticid genera. Along with the naticids and the large ecphoras, numerous smaller drilling muricids were also present in these ecosystems, and these competed with the smaller naticids for bivalve prey. Some of these, which are illustrated in the following iconography, included the genera *Lirosoma*, *Scalaspira*, *Patuxentrophon*, *Chesatrophon*, *Stephanosalpinx*, and *Mariasalpinx*. Along with the larger ecphora ocenebrine muricids that were discussed earlier, these constitute one of the largest assemblages of carnivorous gastropods found in any known marine ecosystem in the American fossil record.

Iconography of Eastern United States Miocene Mollusks

Figure 6.4 Gastropods of the Aquitanian Silverdale Subsea, Albemarle Sea.
A, B= *Atraktus stromboidella* (Kellum, 1926), length 62 mm; **C=** *Cymatophos kellumi* Petuch, 2004, length 46mm; **D=** *Spinifulgur onslowensis* (Kellum, 1926), length 42 mm; **E=** *Gradiconus postalveatus* (Kellum, 1926), length 39 mm; **F=** *Calvertitella fuerta* (Kellum, 1926), length 64 mm; **G=** *Tritonopsis gilletti* (Richards, 1943), length 25 mm; **H=** *Tritonopsis biconica* (Dall, 1915), length 19 mm. All specimens were collected in the Silverdale Marl Company quarry, Silverdale, Onslow County, North Carolina, in the Haywood Landing Member of the Belgrade Formation, Aquitanian Miocene. *Tritonopsis biconica* is also found in the contemporaneous Tampa Member of the Arcadia Formation in southwestern Florida (in the Tampa Subsea of the Okeechobean Sea). This Silverdale fauna is also present, in an impoverished form, in the Old Church Formation of Maryland (Old Church Subsea of the Salisbury Sea).

Figure 6.5 Ecphoras of the Langhian Pamlico Subsea, Albemarle Sea.
A= *Siphoecphora aurora* (Wilson, 1987), length 70 mm; **B, C**= *Ecphorosycon pamlico* (Wilson, 1987), length 74 mm; **D**= *Trisecphora (Matoaka) carolinensis* (Petuch, 1988), length 30 mm; **E**= *Ecphora wardi* Petuch, 1988, length 23 mm; **F**= *Trisecphora (Eccentrella) chamnessi* Petuch, 1988, length 49 mm. Collected in the PotashCorps Mine (old Lee Creek Mine), Aurora, Beaufort County, North Carolina, in the Bonnerton Member of the Pungo River Formation, Langhian Miocene. *Ecphorosycon pamlico* and *Ecphora wardi* are also found in the contemporaneous Plum Point Member of the Calvert Formation of Maryland (in the Calvert Subsea of the Salisbury Sea).

Figure 6.6 Mollusks of the Langhian Pamlico Subsea, Albemarle Sea.
A= *Pecten maclellani* Gibson, 1987, length 90 mm; **B=** *Chesathais pungoensis* Petuch and Berschauer, 2023, holotype, length 21 mm; **C=** *Cirsotrema calvertensis* (Martin, 1904), length 28 mm; **D=** *Stenorhytis pachypleura* (Conrad, 1841), length 15 mm; **E=** *Amaea prunicola* (Martin, 1904), length 20 mm; **F, G=** *Siphoecphora gerrardi* Petuch and Berschauer, 2023, holotype, length 31 mm; **H=** *Trisecphora (Recurvecphora) recurvicostata* Petuch and Berschauer, 2023, holotype, length 21 mm. Collected in the PotashCorps Mine (old Lee Creek Mine), Aurora, Beaufort County, North Carolina, in the Bonnerton Member of the Pungo River Formation, Langhian Miocene. *Cirsotrema calvertensis*, *Stenorhytis pachypleura*, and *Amaea prunicola* are also found in the contemporaneous Plum Point Member of the Calvert Formation of Maryland (in the Calvert Subsea of the Salisbury Sea).

Figure 6.7 Mollusks of the Langhian Calvert Subsea, Salisbury Sea.
A= *Xenophora lindae* Petuch and Drolshagen, 2010, width 30 mm; **B**= *Scaphellopsis coronaspira* Petuch, 1988, length 35 mm; **C**= *Volutifusus caricelloides* Petuch, 1988, length 65 mm; **D**= *Calverturris bellacrenata* (Conrad, 1841), length 28 mm; **E**= *Conradconfusus devexus* (Conrad, 1843), length 62 mm; **F**= *Torcula exaltata* (Conrad, 1841), length 93 mm; **G**= *Chesapecten coccymelus* (Dall, 1898), length 53 mm; **H**= *Torcula cumberlandiana* (Conrad, 1863), length 90 mm. Collected in cliffs at Plum Point, Calvert County, Maryland, in the Plum Point Member of the Calvert Formation (Shattuck Zone 10). Northern Calvert Cliffs of western Chesapeake Bay, Langhian Miocene.

Figure 6.8 Mollusks of the Langhian Calvert Subsea, Salisbury Sea.
A= *Ptychosalpinx lienosus* (Conrad, 1843), length 50 mm; **B**= *Americoliva harrisi* (Martin, 1904), length 24 mm; **C**= *Calvertitella indenta calvertensis* (Oleksyshyn, 1959), length 41 mm; **D**= *Conradconfusus migrans* (Conrad, 1843), length 65 mm; **E**= *Glossus markoei* (Conrad, 1842), left valve, length 51 mm; **F**= *Ficus harrisi* (Martin, 1904), length 29 mm (broken); **G**= *Carinorbis dalli* (Whitfield, 1894), length 13 mm; **H**= *Ecphorosycon kalyx* (Petuch, 1988), length 68 mm. Collected in cliffs at Plum Point, Calvert County, Maryland, in the Plum Point Member of the Calvert Formation (Shattuck Zones 10 and 12), northern Calvert Cliffs of western Chesapeake Bay, Langhian Miocene.

Figure 6.9 Ecphoras of the Langhian Calvert Subsea, Salisbury Sea.
A= *Chesathais whitfieldi* Petuch, 1988, length 57 mm; **B, C**= *Trisecphora (Matoaka) prunicola* (Petuch, 1988), length 33 mm; **D, E**= *Trisecphora (Recurvecphora) schmidti* (Petuch, 1988), length 30 mm; **F**= *Trisecphora tricostata* (Martin, 1904), length 58 mm; **G**= *Chesathais ecclesiasticus* (Dall, 1915), length 50 mm. Collected in cliffs at Plum Point, Calvert County, Maryland, in the Plum Point Member of the Calvert Formation (Shattuck Zone 10), northern Calvert Cliffs of western Chesapeake Bay, Langhian Miocene.

Figure 6.10 The Eccentric Ecphora of the Langhian Calvert Subsea.
A, B= *Trisecphora (Eccentrella) eccentrica* (Petuch, 1988), length 56 mm; **C**= *Trisecphora (Eccentrella) eccentrica* (Petuch, 1988), length 74 mm, embedded in a block of clayey-sand matrix containing characteristic Plum Point Member mollusks such as the naticid *Euspirella tuomeyi* (Whitfield, 1894) and *Carinorbis dalli* (Whitfield, 1894). All specimens were collected in cliffs at Plum Point, Calvert County, Maryland, in the Plum Point Member of the Calvert Formation (Shattuck Zone 10), in the northern Calvert Cliffs of western Chesapeake Bay, Langhian Miocene. The uncoiling of the body whorl of this species makes it one of the most fascinating and unusual muricids in the entire American fossil record.

Figure 6.11 Ecphoras of the Late Langhian Calvert Subsea, Salisbury Sea.
A= *Planecphora turneri* (Petuch, 1992), length 40 mm; **B**= *Trisecphora (Matoaka) bartoni* Petuch and Drolshagen, 2010, length 30 mm; **C, D**= *Ecphora mattinglyi* Petuch, 2004, length 88 mm; **E**= *Ecphora (Powhatan) calvertensis* Petuch, 1988, length 43 mm; **F**= *Trisecphora (Recurvecphora) martini* (Petuch, 1988), length 25 mm; **G**= *Trisecphora governorensis* Petuch and Berschauer, 2023, length 32 mm; **H**= *Trisecphora scientistensis* (Petuch, 1992), length 42 mm. All specimens were collected in the Calvert Beach Member of the Calvert Formation (Shattuck Zones 14 and 16), along a stretch of the Calvert County, Maryland coast extending from Scientist's Cliffs south to St. Leonards.

Figure 6.12 Mollusks of the Serravallian Patuxent Subsea, Salisbury Sea.
A= *Patuxentrophon patuxentensis* (Martin, 1904), length 18 mm; **B=** *Volutifusus conradianus* (Martin, 1904), length 57 mm ("*V. virginianus* Dall", a nomen nudum, is a synonym); **C=** *Sycopsis lindae* Petuch, 1988, length 67 mm; **D=** *Conradconfusus patuxentensis* (Petuch, 1993), length 74 mm; **E=** *Ventrila biplicifera* (Conrad, 1841), length 42 mm; **F=** *Ecphorosycon lindajoyceae* Petuch, 1993, length 69 mm; **G, H=** *Stephanosalpinx candelabra* Petuch, 1988, length 18 mm (incomplete). Collected in the St. Leonard Member (Shattuck Zone 17) and Drumcliff Member (Shattuck Zone 18) of the Choptank Formation at Matoaka Cottages, St. Leonards, Calvert County, Maryland and along the Patuxent River, St. Mary's County, Maryland, Serravallian Miocene.

Figure 6.13 Ecphoras of the Serravallian Patuxent Subsea, Salisbury Sea.
A= *Planecphora delicata* (Petuch, 1988), length 39 mm; **B=** *Planecphora choptankensis* (Petuch, 1988), length 62 mm; **C=** *Ecphora (Rhombecphora) rikeri* Petuch, 1988, length 80 mm; **D=** *Ecphora sandgatesensis* Petuch, 1988, length 40 mm; **E=** *Trisecphora (Matoaka) patuxentia* (Petuch, 1988), length 37 mm; **F=** *Ecphora (Powhatan) williamsi* Ward and Gilinsky, 1988, length 57 mm; **G=** *Chesathais lindae* Petuch, 1988, length 36 mm; **H=** *Chesathais drumcliffensis* Petuch, 1988, length 32 mm. Collected in the St. Leonard Member (Shattuck Zone 17), Drumcliff Member (Shattuck Zone 18), and Boston Cliffs Member (Shattuck Zone 19) of the Choptank Formation at Matoaka Cottages, St. Leonards, Calvert County, Maryland and along the Patuxent River, St. Mary's County, Maryland, Serravallian Miocene.

Figure 6.14 Mollusks of the late Serravallian St. Mary's Subsea, Salisbury Sea.
A, B= *Ecphora (Powhatan) asheri* Petuch, 1988, length 75 mm; **C, D**=*Ecphora (Powhatan) conoyensis* Petuch, 2004, length 51 mm (Conoy Member of the Choptank Formation only); **E**= *Scalaspira vokesae* Petuch, 1988, length 18 mm; **F**= *Dallarca covepointensis* Petuch and Drolshagen, 2010, length 48 mm; **G**= *Marianarona lindae* Petuch and Drolshagen, 2010, length 17 mm; **H**= *Mariacolpus covepointensis* Petuch, 2004, length 33 mm. All, except for Figures C and D, were collected in the lower bed of the Little Cove Point Member (Shattuck Zone 21) of the St. Mary's Formation at Cove Point , Calvert County, Maryland, late Serravallian Miocene.

Figure 6.15 Gastropods of the late Serravallian St. Mary's Subsea, Salisbury Sea.
A= *Mariasalpinx bartoni* Petuch and Drolshagen, 2010, length 68mm; **B**= *Turrifulgur covepointensis* Petuch, 2004, length 68 mm; **C**= *Conradconfusus chesapeakensis* (Petuch, 1988), length 80 mm; **D**= *Sycopsis carinatum* (Conrad, 1862), length 69 mm; **E**= *Coronafulgur chesapeakensis* (Petuch, 1988), length 95 mm; **F**= *Sycofulgur martini* (Petuch and Drolshagen, 2010), length 178 mm; **G**= *Ecphora (Powhatan) germonae* Ward and Gilinsky, 1988, length 83 mm; **H**= *Mercenaria lindae* Petuch and Drolshagen, 2011, length 65 mm. Collected in the upper bed of the Little Cove Point Member (Shattuck Zone 22) at Cove Point, Calvert County, Maryland, early Tortonian Miocene.

Figure 6.16 Gastropods of the Tortonian St. Mary's Subsea, Salisbury Sea.
A= *Mariacolpus lindajoyceae* Petuch, 2004, length 38 mm; **B=** *Chesatrophon harasewychi* (Petuch, 1988), length 16 mm; **C=** *Chesatrophon lindae* (Petuch, 1988), length 15 mm; **D=** *Lirosoma mariana* Petuch, 1988, length 25 mm; **E=** *Scalaspira harasewychi* Petuch, 1988, length 15 mm; **F=** *Stenorhytis expansa* (Conrad, 1842), length 23 mm; **G=** *Chesatrophon laevis* (Martin, 1904), length 16 mm; **H=** *Poliniciella marylandica* Petuch, 1988, length 24 mm; **I=** *Chesatrophon chesapeakensis* (Martin, 1904), length 5 mm. All specimens were collected in low cliffs along the St. Mary's River at Chancellor Point, St. Mary's City, St. Mary's County, Maryland, in the Windmill Point Member of the St. Mary's Formation (Shattuck Zone 24), Tortonian Miocene.

Figure 6.17 Gastropods of the Tortonian St. Mary's Subsea, Salisbury Sea.
A, B= *Eudolium marylandicum* Petuch, 1988, length 44 mm; **C, D=** *Ecphora (Powhatan) gardnerae* Wilson, 1987, length 96 mm; **E=** *Mariafusus marylandicus* (Martin, 1904), length 60 mm; **F=** *Chesapecten santamaria* (H. Tucker, 1934), length 135 mm; **G=** *Coronafulgur coronatum* (Conrad, 1840), length 98 mm; **H=** *Ptychosalpinx lindae* Petuch, 1988, length 26 mm. All specimens were collected in low cliffs along the St. Mary's River at Chancellor Point, St. Mary's City, St. Mary's County, Maryland, in the Windmill Point Member of the St. Mary's Formation (Shattuck Zone 24), Tortonian Miocene.

Figure 6.18 Gastropods of the Tortonian St. Mary's Subsea, Salisbury Sea.
A= *Conradconfusus parilis* (Conrad, 1832), length 106 mm; **B=** *Turrifulgur fusiforme* (Conrad, 1840), length 69 mm; **C=** *Sycopsis tuberculatum* (Conrad, 1840), length 58 mm; **D=** *Sycofulgur rugosum* (Conrad, 1843), length 72 mm; **E=** *Hesperisternia petuchi* Vermeij, 2006, length 23 mm; **F=** *Mariasalpinx subrusticus* (d'Orbigny, 1852), length 25 mm; **G=** *Ptychosalpinx pustulosus* Petuch, 1988, length 20 mm; **H=** *Ilyanssa peralta* (Conrad, 1868), length 22 mm. All specimens were collected in low cliffs along the St. Mary's River at Chancellor Point, St. Mary's City, St. Mary's County, Maryland, in the Windmill Point Member of the St. Mary's Formation (Shattuck Zone 24), Tortonian Miocene.

Figure 6.19 Volutes of the Tortonian St. Mary's Subsea, Salisbury Sea.
A= *Volutifusus marylandicus* Petuch, 1988, holotype, length 144 mm; **B, C**= *Volutifusus marylandicus* Petuch, 1988, length 146 mm; **D**= *Volutifusus asheri* Petuch, 1988, length 74 mm; **E, F**= *Volutifusus acus* Petuch, 1988, length 96 mm; **G, H**= *Volutifusus mutabilis* (Conrad, 1843), length 99 mm; **I**= *Scaphellopsis solitaria* (Conrad, 1830), length 26 mm. All specimens were collected in low cliffs along the St. Mary's River at Chancellor Point, St. Mary's City, St. Mary's County, Maryland, in the Windmill Point Member of the St. Mary's Formation (Shattuck Zone 24), Tortonian Miocene.

Figure 6.20 Cone Shells of the Tortonian St. Mary's Subsea, Salisbury Sea.
A, B= *Gradiconus asheri* (Petuch, 1988), length 42 mm; **C, D**= *Gradiconus sanctaemariae* (Petuch, 1988), length 40 mm; **E, F**= *Gradiconus deluvianus* (Green, 1830), length 58 mm; **G**= *Gradiconus deluvianus* (Green, 1830), length 70 mm. All specimens were collected in low cliffs along the St. Mary's River at Chancellor Point, St. Mary's City, St. Mary's County, Maryland, in the Windmill Point Member of the St. Mary's Formation (Shattuck Zone 24), Tortonian Miocene.

Figure 6.21 Gastropods of the Tortonian St. Mary's Subsea, Salisbury Sea.
A= *Bulliopsis quadrata* (Conrad, 1830), length 23 mm; **B**= *Mariaturricula biscatenaria* (Conrad, 1830), length 36 mm; **C**= *Chesasyrinx rotifera* (Conrad, 1830), length 23 mm; **D**= *Chesasyrinx mariana* (Martin, 1904), length 26 mm; **E**= *Cymatosyrinx limatula* (Conrad, 1830), length 24 mm; **F**= *Marianarona alternata* (Conrad, 1834), length 22 mm; **G**= *Marianarona marylandica* Petuch, 1988, length 12 mm; **H**= *Laevihastula sublirata* (Conrad, 1863), length 29 mm; **I**= *Ecphora (Powhatan) gardnerae jeffersoni* Petuch and Berschauer, 2026, holotype, length 92 mm, collected in a blue clay layer filled with gypsum crystals, along the Rosecroft Cliffs, St. Mary's County, Maryland; in an unnamed lowest bed of the Eastover Formation. All the other specimens were collected in low cliffs along the St. Mary's River at Chancellors Point, St. Mary's City, St. Mary's County, Maryland, in the Windmill Point Member of the St. Mary's Formation (Shattuck Zone 24), Tortonian Miocene.

Figure 6.22 Mollusks of the Late Tortonian Rappahannock Subsea, Albemarle Sea.
A, B= *Ecphora (Powhatan) whiteoakensis* Ward and Gilinsky, 1988, length 127 mm; **C=** *Volutifusus rappahannockensis* Petuch and Drolshagen, 2010, length 89 mm; **D=** *Astarte lindae* Petuch and Drolshagen, 2010, length 21 mm; **E=** *Mercenaria gardnerae* Petuch and Drolshagen, 2010, length 68 mm; **F=** *Chesapecten ceccae* Ward, 1992, length 174 mm; **G=** *Americoliva eborea* (Conrad, 1862), length 35 mm. Collected at the base of a low cliff along Urbanna Creek, Urbanna, Middlesex County, Virginia, in the Claremont Manor Member of the Eastover Formation, late Tortonian Miocene.

Figure 6.23 Gastropods of the Messinian Rappahannock Subsea, Albemarle Sea.
A= *Coronafulgur kendrewi* Petuch, 2004, length 122 mm; **B, C=** *Americoliva idonea* (Conrad, 1839), length 42 mm; **D=** *Ecphora kochi* Ward and Gilinsky, 1988, length 92 mm; **E=** *Gradiconus spenceri* (Ward, 1992), length 33 mm; **F=** *Neverita (Glossaulax) vautrotorum* Petuch and Drolshagen, 2010, length 39 mm; **G=** *Mariacolpus plebeius* (Say, 1824), length 41 mm; **H=** *Ecphora (Powhatan) grulkei* Petuch and Berschauer, 2023, holotype, length 76 mm. The ecphoras were collected in the Lee Creek Phosphate Mine, Aurora, North Carolina. All the other species were collected in low cliffs along the Rappahannock River, between Bowler's Wharf and Urbanna, Middlesex County, Virginia, in the Cobham Bay Member of the Eastover Formation, early Messinian Miocene.

Figure 6.24 Mollusks of the Messinian Rappahannock Subsea, Albemarle Sea.
A, B= *Crepidula (Bostrycapulus) bonnieae* Petuch and Drolshagen, 2010, length 23 mm; **C, D=** *Euspirella permeliae* Petuch and Drolshagen, 2010, length 14 mm (note the distinctly angled shoulder); **E=** *Mansfieldostrea geraldjohnsoni* (Ward, 1992), lower valve, length 116 mm; **F=** *Astarte rappahannockensis* Gardner, 1944, length 24 mm; **G=** *Chesapecten middlesexensis* (Mansfield, 1936), length 158 mm; **H=** *Cyclocardia vautrotorum* Petuch and Drolshagen, 2010, length 25 mm. Collected in low cliffs along the Rappahannock River, between Bowler's Wharf and Urbanna, Middlesex County, Virginia, in the Cobham Bay Member of the Eastover Formation, early Messinian Miocene.

CHAPTER 7. Pliocene Molluscan Faunas of the Eastern United States

The severe worldwide cooling seen at the end of the Messinian Miocene, which was caused by the Kara-Kul Asteroid Impact, had an immediate impact on the marine environments along the eastern United States. All the tropical marine molluscan genera that inhabited the late Messinian Rappahannock Subsea, such as *Gradiconus*, *Americoliva*, and *Volutifusus*, abruptly disappeared from the local fossil record but survived farther south in the Okeechobean Sea of Florida (Petuch, 2004; Petuch and Berschauer, 2022). The entire Transmarian Province ceased to exist at this time, with only isolated pockets of relictual taxa surviving into the Zanclean Age of the early Pliocene. One of these surviving relict pockets of older, archaic genera is seen in the northern Albemarle Sea, in the fauna of the lowest beds of the Yorktown Formation (Mansfield Zone 1). Here, prominent Transmarian genera such as the muricids *Lirosoma*, *Scalaspira*, *Mariasalpinx*, and *Chesatrophon*, the volutes *Scaphellopsis* and *Megaptygma*, the terebrid *Laevihastula*, and the drilliid *Sediliopsis* persisted into an entirely new paleoprovince that was just beginning to evolve at that time (examples of these genera are shown here in the Iconography at the end of this chapter).

The Transmarian relictual elements living in the northern Rappahannock Subsea during Zanclean time were quickly absorbed into a new hybrid paleoprovince, the **Caloosahatchian Province** (named for the Caloosahatchee Formation of Gelasian southern Florida, which contains the classic molluscan fauna; see Petuch, 1982; 1988; 2004). The core fauna of the Caloosahatchian Province evolved within the **Okeechobean Sea** and **Everglades Pseudoatoll** of Pliocene Florida and contained characteristic endemic genera such as the cypraeid *Akleistostoma*, the cone shell *Contraconus*, the volutid *Scaphella*, the fasciolariids *Cinctura* and *Triplofusus*, and the busyconids *Busycon*, *Fulguropsis* and *Sinistrofulgur* (examples of these Caloosahatchian genera are shown in the Iconography at the end of this chapter; see Petuch and Berschauer, 2022). All of these genera made their first appearance in the Pliocene fossil beds of Florida and the eastern United States. Many of these Caloosahatchian genera persisted into the Pleistocene and eventually lived on to become components of the modern Carolinian Province, surviving the numerous cold times of the late Pleistocene. At its peak development during the late Piacenzian Age, the Caloosahatchian Province extended from Virginia south to Florida and the entire eastern half of the Gulf of Mexico.

The Pliocene Charleston and Pliocene Albemarle Seas

By the beginning of the Zanclean Age of the early Pliocene, the axis of deposition within the Albemarle Sea had moved southward, encompassing what is today the Pamlico Sound and Cape Hatteras areas of North Carolina. At this time, a new paleosea was beginning to form in what had been coastal marshlands and lagoons during Messinian time. This new oceanic system was named the **Charleston Sea** (named for Charleston, South Carolina; Figure 7.1), and it was separated from the Albemarle Sea to the north by the Cape Fear highlands structural arch (Petuch, 2004:136). Along its southern half, the Charleston Sea connected to a large series of coastal lagoons and estuaries, the **Satilla Lagoon System** (Petuch, 2004:152), which extended southward along eastern Florida and into the Okeechobean Sea of the Everglades region. The Satilla Lagoon section of the Charleston Sea shared many mollusks with the extensive mangrove estuaries of the

DOI: 10.1201/9781003732983-7

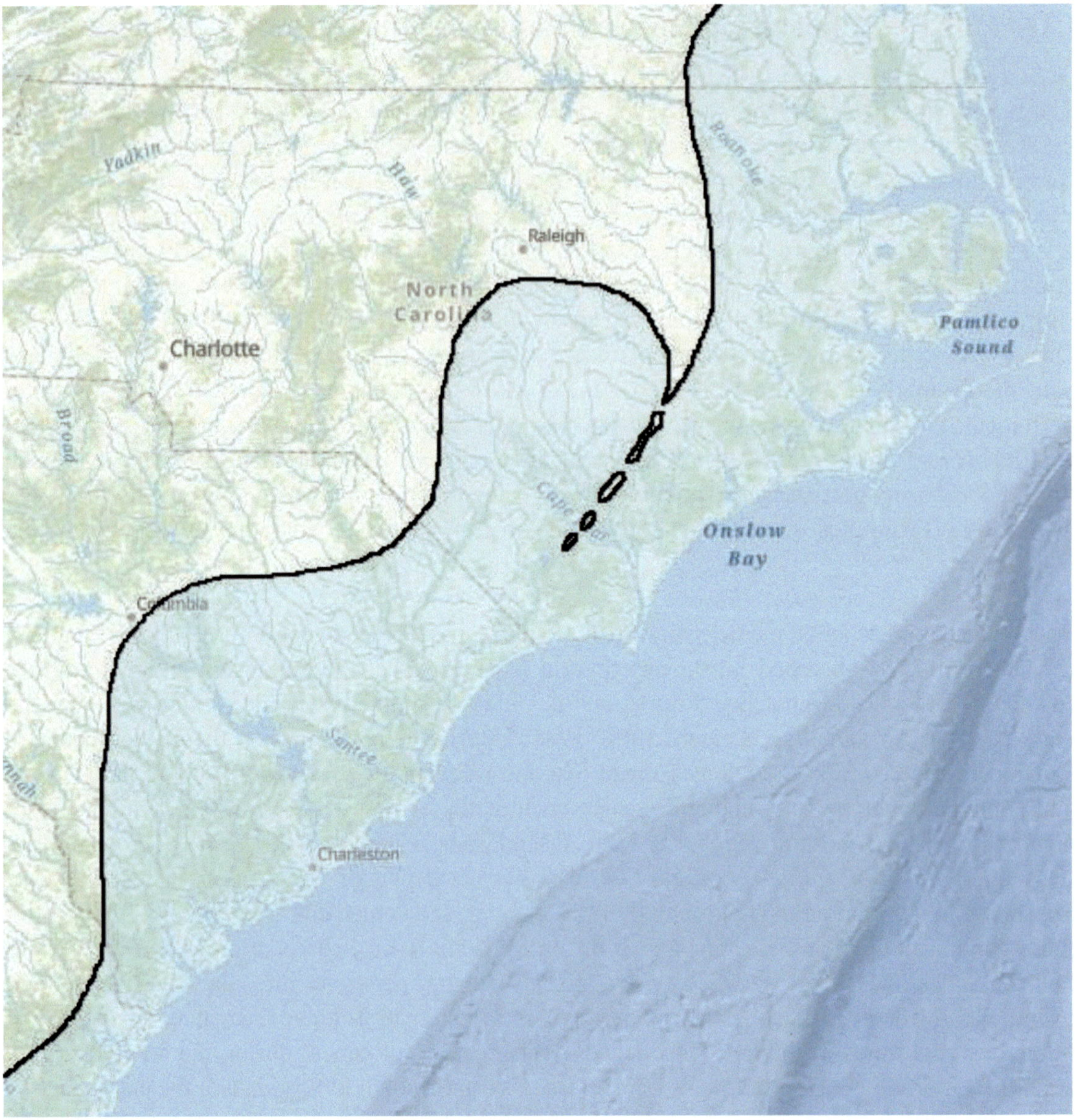

Figure 7.1 Map of the Pliocene Charleston Sea during its maximum development in the Piacenzian Age. The black line denotes the outline of the paleosea superimposed upon a recent map of the coast of the Carolinas.

Okeechobean Sea, including gastropods such as the potamids *Potamides matsoni* and *Potamides gracilior*, the neritids *Smaragdia merida* and *Neritina sphaerica*, and the mactrid bivalve *Mulinia sapotilla*. (For a detailed review of the oceanography of the Charleston Sea, see Petuch, 2004). Farther north, the **Pliocene Albemarle Sea** (Figure 7.2) had expanded into a larger oceanic system that extended farther southward along the coast, with its geographic center located westward of the present-day Pamlico Sound of North Carolina. A detailed overview of the geomorphology, island chains, and resident ecosystems of the Pliocene Albemarle Sea was given by Petuch (2004:144-152).

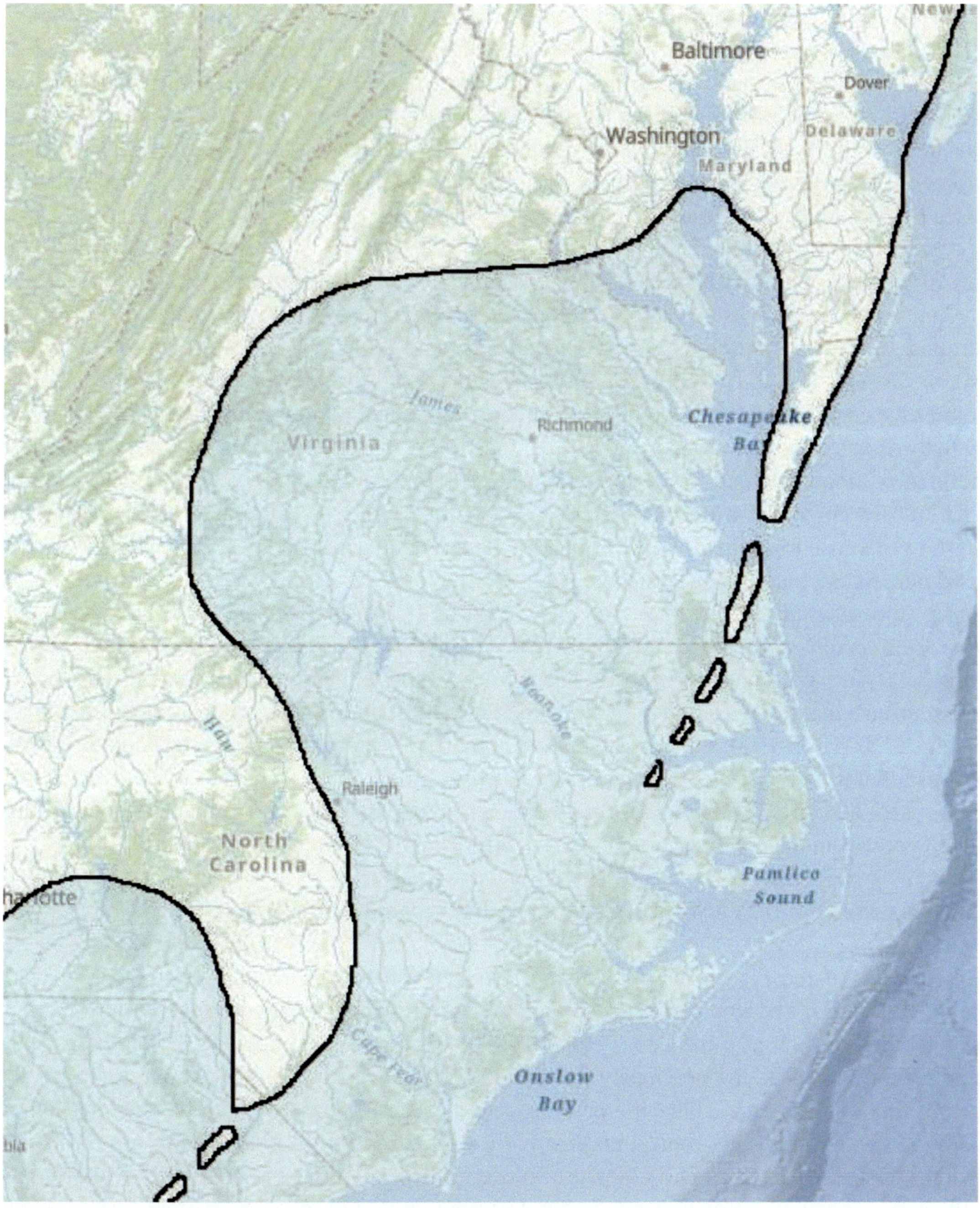

Figure 7.2 Map of the Pliocene Albemarle Sea during its maximum development in the Piacenzian Age. The black line denotes the outline of the paleosea superimposed upon a recent map of the coast of North Carolina and Virginia.

The Zanclean and earliest Piacenzian Subseas of the East Coast

During the early Pliocene (Zanclean and earliest Piacenzian Ages), the structural basins of the Charleston and Pliocene Albemarle Seas were separated from each other by the highlands of the Cape Fear Arch, a high ridge that ran perpendicular to the main axis of the Appalachian Mountains. This geomorphological feature extended farther out into the open Atlantic Ocean than present-day Cape Fear and may have been a remnant of a debris field that was produced by the

Chesapeake Bay Asteroid Impact. Although sharing many species in common, each of the two isolated paleoseas evolved its own separate molluscan fauna, with many endemic taxa, and the fossil beds from both oceanic basins look very different. The Zanclean beds of the Charleston Sea were deposited in the **Santee Subsea** (named for the Santee River of South Carolina, which has exposures of the typical fauna) while chronologically equivalent beds deposited within the Pliocene Albemarle Sea were deposited within the larger and better developed **Williamsburg Subsea** (named for beds exposed at Williamsburg, Virginia).

Many of the wide-ranging Zanclean-earliest Piacenzian mollusks occurred within both paleoseas, including bivalves such as the large scallop *Chesapecten jeffersonius*, and classic East Coast gastropods such as the muricids *Pterorhytis umbrifer* and *Ecphora quadricostata*, the busyconid *Busycon filosum*, and the volutid *Volutifusus emmonsi* (all shown here on Figure 7.3). These ubiquitous species have also been collected in contemporaneous formations throughout Florida to the south (Petuch and Berschauer, 2022), demonstrating that there was a well-established and widespread Caloosahatchian Province fauna at this time. Being geographically farther south, the Santee Subsea contained a warmer water fauna, with several tropical genera that were missing from the Williamsburg Subsea. Some of these, which lived during the latest Zanclean and earliest Piacenzian Ages, included gastropods such as the cowrie *Akleistostoma (Ingramicypraea) pilsbryi*, the muricids *Ecphora roxaneae dalli* and *Latecphora violetae*, the busyconid *Fulguropsis carolinensis*, and the pleioptygmatid miter *Pleioptygma carolinensis* (shown here on Figure 7.4). The Williamsburg Subsea had cooler water temperatures, primarily due to cold countercurrents that flowed southward along the East Coast at this time (the Narragansett Current; see Petuch, 2004:135-137). This colder marine climate is seen in the impoverished nature of the molluscan fauna, which was dominated by bivalves such as the pholadomyid *Margaritaria abrupta* and the ostreid *Mansfieldostrea compressirostra* and gastropods that included a large fauna of naticids in the cold-water genus *Euspira* and small *Solariella* trochoideans.

The Piacenzian Duplin, Yorktown, Croatan and Waccamaw Subseas

The beginning of the Piacenzian Age of the Pliocene ushered in a long period of very warm climatic conditions, and tropical warm-temperate environments extended up the eastern coast of the United States as far as Cape Cod (Petuch, 1988b; 2004). At this time, the Okeechobean Sea of southern Florida housed the previously mentioned Everglades Pseudoatoll, the largest coral reef structure found in the entire Pliocene or Pleistocene western Atlantic. Many of the species that had evolved within this high tropical marine world migrated northward during this time of Global Warming (discussed in Chapter 5). Some of these tropical migrants, whose oldest examples are found in the Tamiami Formation of southern Florida, included gastropods such as the cowrie *Akleistostoma carolinensis*, the muricid *Phyllonotus globosus*, the busyconid *Sinistropfulgur contrarium*, the cone shell *Contraconus adversarius*, the olivid *Americoliva carolinensis*, and the terebrid *Cinguloterebra unilineata*, and bivalves such as the pectinid *Nodipecten peedeensis*. These classic widespread Caloosahatchian Province taxa are shown here in the Iconography at the end of this chapter. By the mid-Pliocene, these southern migrants became established within the **Duplin Subsea** (named for the Duplin Formation of North Carolina) of the Charleston Sea and the **Yorktown Subsea** (named for the Yorktown Formation of Virginia) of the Pliocene Albemarle Sea (Petuch, 2004).

Although sharing numerous typical Pliocene Caloosahatchian genera and species between them, both paleoseas also evolved a number of distinctive endemic species. Within the Duplin Subsea, geographically restricted genera and species, such as fasciolariid *Terebraspira elegans* (Figure 7.7A) and the heavily-ribbed volute *Bruggemania precursor* (Figure 7.7B, C), made their first appearance at this time and were prominent predatory gastropods in the shallow-water ecosystems of the Duplin Formation. Farther north within the Yorktown Subsea, the giant predatory volute, *Megaptygma sinuosa* (a juvenile specimen is shown here on Figure 7.11F), dominated the nearshore ecosystems and reached lengths of over 225 mm. Since this outsized scaphelline volute is extremely fragile and thin shelled, it is known only from a few broken large specimens, incomplete spire whorls, and numerous fragments of the body whorl. An intact specimen of a full-grown adult *Megaptygma* would be one of the most spectacular American fossil mollusks and a true marvel from the ecosystems of the Yorktown Subsea.

During the late Piacenzian Pliocene, the marine climate began to cool and sea level dropped by at least 30 m worldwide. Because of this eustatic low and increased sedimentary infilling, the Albemarle and Charleston Seas shrank in size and became much smaller basins, which is less than half of its previous size and volume. This reduction of size and bathymetric ranges, coupled with colder water temperatures, decimated many of the well-established older ecosystems, leading to the extinctions of numerous prominent Caloosahatchian molluscan genera. Primary among these was the complete removal of the ecphora muricids from the fossil record, with the genera *Ecphora*, *Planecphora*, *Latecphora*, and *Globecphora* abruptly disappearing before the deposition of latest Pliocene formations, such as the Chowan River Formation within the Albemarle Sea and the lower beds of the Waccamaw Formation within the Charleston Sea. The abrupt disappearance of the ecphoras presaged the extinction of many other important mollusks which had dominated the Miocene and Pliocene seas.

After the late Pliocene extinction event and the corresponding diminishing of the oceanic basin, the Duplin and Yorktown Subseas were permanently altered as marine systems and were replaced by two new subseas: the **Croatan Subsea** in the Albemarle Sea and the **Waccamaw Subsea** in the Charleston Sea (see Petuch, 2004 for details on the geomorphology and oceanography of these subseas). The last remnants of the rich early and mid-Pliocene molluscan faunas of the Caloosahatchian Province are shown here on Figures 7.12, 7.13, and 7.14. Some of the gastropod and bivalve genera that became extinct at this time are listed here:

EAST COAST GENERA THAT BECAME EXTINCT BY THE LATE PLIOCENE
Gastropoda
Cypraeidae
Akleistostoma Gardner, 1948
Akleistostoma (Ingramicypraea) Petuch and Drolshagen, 2011
Muricidae
Chesatrophon Petuch, 1988
Ecphora Conrad, 1843
Ecphora (Powhatan) Petuch and Berschauer, 2023
Globecphora Petuch, 1994
Latecphora Petuch, 1988
Lirosoma Conrad, 1862
Planecphora Petuch, 2004

Scalaspira Conrad, 1862
Fasciolariidae
Terebraspira Emmons, 1858
Volutidae
Megaptygma Gabb, 1862
Bruggemania Petuch and Berschauer, 2026
Scaphellopsis Petuch, 1988
Terebridae
Laevihastula Petuch, 1988
Drilliidae
Sediliopsis Petuch, 1988
Bivalvia
Pectinidae
Chesapecten Ward and Blackwelder, 1975
Ostreidae
Mansfieldostrea Petuch and Drolshagen, 2009
Cardiidae
Chesacardium Ward, 1992

Close-up photo of an exposure of the Yorktown Formation in the old Lee Creek Phosphate Mine at Aurora, Beaufort County, North Carolina, showing a small *Ecphora (Powhatan) pachycostata* Petuch, 1988 embedded in the phosphatic sediments of the Sunken Meadow Member (Zone 1). The senior author's favorite geological hammer is shown for scale.

Iconography of Eastern United States Pliocene Mollusks

Figure 7.3 Mollusks of the Zanclean Williamsburg Subsea, Albemarle Sea.
A= *Ecphora (Powhatan) pachycostata* Petuch, 1988, length 48 mm; **B, C**= *Pterorhytis umbrifer* (Conrad, 1832), length 37 mm; **D**= *Ecphora quadricostata* (Say, 1824), length 41 mm; **E**= *Chesapecten jeffersonius* (Say, 1824), length 124 mm; **F**= *Volutifusus emmonsi* Petuch, 1994, length 110 mm; **G**= *Busycon filosum* (Conrad, 1862), length 190 mm. These chronostratigraphic index species are present in all well-preserved Zanclean beds, from southern Florida to North Carolina (Sunken Meadow Member, Mansfield Zone 1, Yorktown Formation and equivalent beds).

Figure 7.4 Gastropods of the Zanclean and earliest Piacenzian Santee Subsea, Charleston Sea. **A=** *Latecphora violetae* (Petuch, 1988), length 98 mm; **B=** *Ecphora roxaneae dalli* Petuch and Berschauer, 2023, length 111 mm; **C=** *Pleioptygma carolinensis* (Tuomey and Holmes, 1858), length 66 mm; **D=** *Globecphora parvicostata* (Pilsbry, 1911), length 57 mm; **E, F=** *Akleistostoma (Ingramicypraea) pilsbryi* (Ingram, 1947), length 26 mm; **G=** *Fulguropsis carolinensis* (Tuomey and Holmes, 1856), length 109 mm. These chronostratigraphic index species are present in all well-preserved Zanclean and earliest Piacenzian Pliocene beds, from southern Florida to North Carolina (Duplin, Goose Creek, and Tamiami Formations).

Figure 7.5 Gastropods of the Zanclean and early Piacenzian Yorktown Subsea, Albemarle Sea. **A**= *Lirosoma sulcosa* (Conrad, 1830), length 34 mm; **B**= *Chesatrophon tetricus* (Conrad, 1832), length 28 mm; **C**= *Scalaspira strumosa* (Conrad, 1830), length 26 mm; **D**= *Aristofusus burnsi* (Dall, 1890), length 87 mm; **E**= *Ptychosalpinx laqueatus* (Conrad, 1832), length 39 mm; **F**= *Americoliva canaliculata* (H. Lea, 1843), length 42 mm; **G**= *Scaphellopsis ricei* (Campbell, 1993), length 34 mm; **H**= *Urosalpinx cannabinus* (Conrad, 1832), length 35 mm. All specimens were collected in the Rushmere Member of the Yorktown Formation (Mansfield Zone 2) at the Lone Star Cement Company quarry, Chuckatuck, Virginia, Piacenzian Pliocene.

Figure 7.6 Mollusks of the Piacenzian Duplin Subsea, Charleston Sea.
A= *Triplofusus duplinensis* (B. Smith, 1940), length 356 mm; **B, C**= *Akleistostoma carolinensis* (Conrad, 1841), length 60 mm; **D**= *Pliculofusus sparrowi* (Emmons, 1858), length 148 mm; **E**= *Mercenaria tridacnoides* (Lamarck, 1818), length 122 mm; **F**= *Cinguloterebra unilineata* (Conrad, 1841), length 86 mm; **G**= *Nodipecten peedeensis* (Tuomey and Holmes, 1856), length 138 mm; **H**= *Scaphella mansfieldi* Petuch, 1994, length 71 mm. These chronostratigraphic index species are present in all well-preserved Piacenzian Pliocene beds, from southern Florida to North Carolina.

Figure 7.7 Gastropods of the Piacenzian Duplin Subsea, Charleston Sea.
A= *Terebraspira elegans* Emmons, 1858, holotype, length 85 mm; **B, C**= *Bruggemania precursor* (Gardner, 1948), length 78 mm; **D, E**= *Latecphora brantleyae* (Petuch, 1987), length 68 mm; **F**= *Ptychosalpinx multirugatus* (Conrad, 1841), length 43 mm; **G**= *Ptychosalpinx bilix* (Conrad, 1843), length 48 mm. The genus *Pliculofusus* Snyder, Vermeij, and Lyons, 2012 appears to be a junior synonym of *Terebraspira*, as the two genera intergrade morphologically.

Figure 7.8 Gastropods of the Piacenzian Duplin Subsea, Charleston Sea.
A= *Sinistrofulgur robesonense* (Gardner, 1948), length 147 mm; **B=** *Americoliva robesonensis* (Gardner, 1948), length 28 mm; **C=** *Volutifusus spengleri* Petuch, 1991, length 160 mm; **D, E=** *Scaphella trenholmii* (Tuomey and Holmes, 1856), length 85 mm; **F=** *Americoliva carolinensis* (Conrad, 1863), length 85 mm; **G=** *Fulguropsis excavatum* (Conrad, 1840), length 102 mm. These chronostratigraphic index species are present in all well-preserved Piacenzian Pliocene beds, from southern Florida to North Carolina.

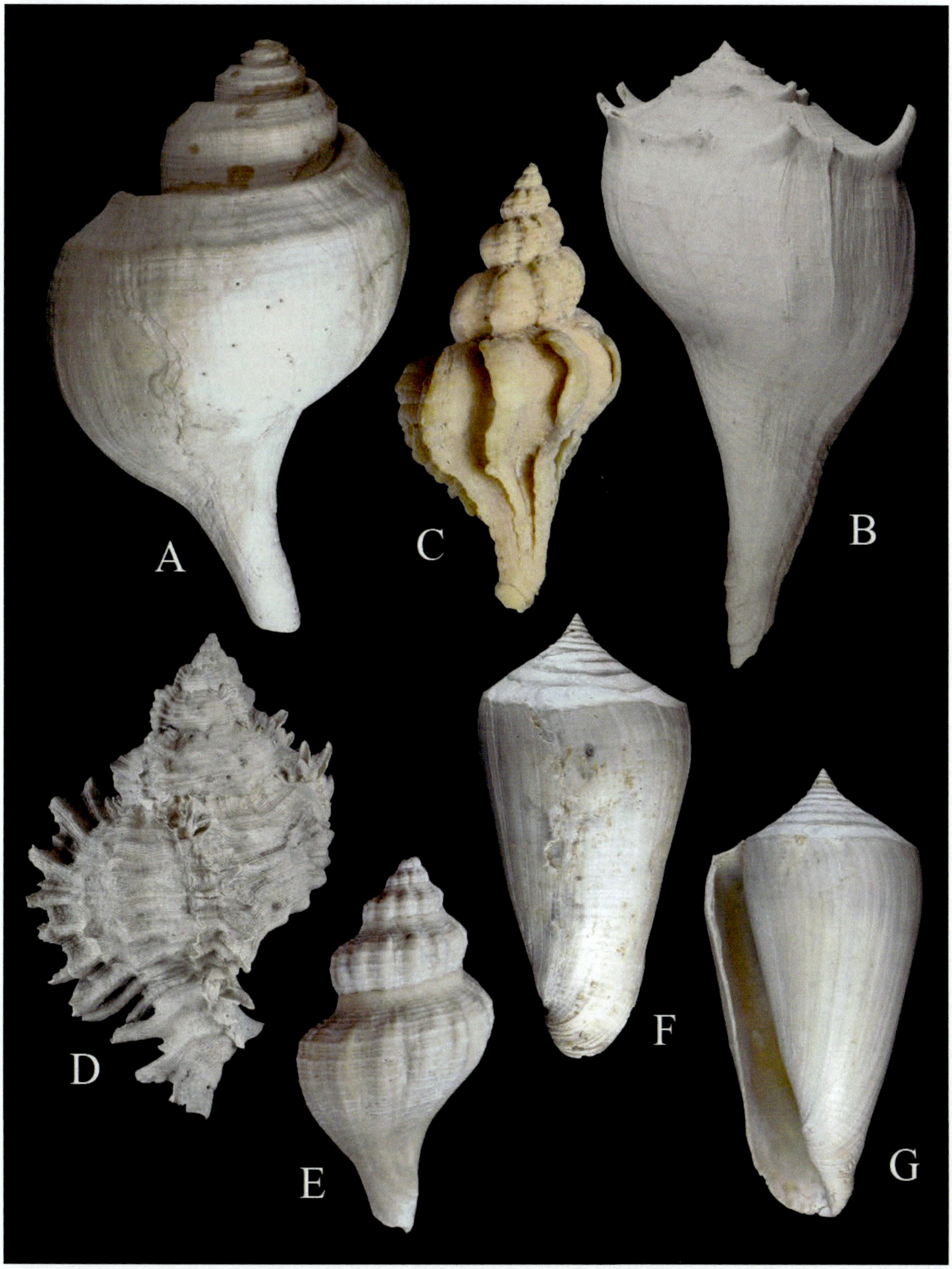

Figure 7.9 Gastropods of the Piacenzian Duplin Subsea, Charleston Sea.
A= *Busycotypus conradi* (Tuomey and Holmes, 1856), length 210 mm; **B=** *Sinistrofulgur contrarium* (Conrad, 1840), length 230 mm; **C=** *Vokesinotus lamellosus* (Emmons, 1858), length 35 mm; **D=** *Phyllonotus globosus* (Emmons, 1858), length 79 mm; **E=** *Mariasalpinx stimpsoni* (Campbell, 1993), length 34 mm; **F, G=** *Contraconus adversarius* (Conrad, 1840), length 98 mm. These chronostratigraphic index species are present in all well-preserved Piacenzian Pliocene beds, from southern Florida to North Carolina.

Figure 7.10 Ecphoras of the Piacenzian Yorktown Subsea, Albemarle Sea.
A, B= *Ecphora quadricostata umbilicata* Martin, 1904 ("Wagner, 1839"), length 68 mm; **C, D=** *Ecphora quadricostata rachelae* Petuch, 1988, length 42 mm; **E, F=** *Globecphora mardieae* Petuch and Berschauer, 2023, length 68 mm; **G, H=** *Planecphora mansfieldi* (Petuch, 1988), length 66 mm. All specimens were collected in the Rushmere Member of the Yorktown Formation (Mansfield Zone 2), in the PotashCorps Mine (old Lee Creek Mine), Aurora, Beaufort County, North Carolina, Piacenzian Pliocene. These are the last ecphoras found in the fossil record of the eastern United States, as the entire complex becomes extinct by the late Piacenzian Age.

Figure 7.11 Mollusks of the Piacenzian Yorktown Subsea, Albemarle Sea.
A= *Busycon maximus* (Conrad, 1839), length 149 mm; **B**= *Busycotypus incile* (Conrad, 1833), length 168 mm; **C, D**= *Volutifusus obtusus* (Emmons, 1858), length 99 mm; **E**= *Ximeniconus marylandicus* (Green, 1830), length 28 mm; **F**= *Megaptygma sinuosa* (Gabb, 1862), juvenile specimen, length 93 mm (fully adult specimens may reach lengths of over 225 mm; see Campbell, 1993); **G**= *Trossulasalpinx trossulus* (Conrad, 1832), length 23 mm; **H**= *Chesapecten septenarius* (Say, 1824), length 82 mm. All specimens were collected in the Rushmere Member of the Yorktown Formation (Mansfield Zone 2) at the Lone Star Cement Company quarry, Chuckatuck, Virginia, Piacenzian Pliocene.

Figure 7.12 Gastropods of the late Piacenzian Croatan Subsea, Albemarle Sea.
A= *Busycon auroraensis* Petuch, 1994, length 165 mm; **B=** *Busycotypus concinnum* (Conrad, 1873), length 283 mm; **C=** *Sinistrofulgur adversarius* (Conrad, 1863), length 205 mm; **D, E=** *Ptychosalpinx chesapeakensis* Campbell, 1993, length 46 mm; **F, G=** *Contraconus petiti* Petuch, 2004, length 87 mm. All specimens were collected in the Chowan River Formation, at the PotashCorps Mine (old Lee Creek Mine), Aurora, Beaufort County, North Carolina, late Piacenzian Pliocene. These species were also present in the Waccamaw Subsea of the Charleston Sea.

Figure 7.13 Gastropods of the late Piacenzian Croatan Subsea, Albemarle Sea.
A, B= *Pterorhytis conradi* (Dall, 1890), length 55 mm; **C**= *Pliculofusus cronleyensis* (Gardner, 1948), length 152 mm; **D, E**= *Volutifusus auroraensis* Petuch, 1994, length 110 mm; **F, G**= *Urosalpinx gilmorei* Petuch, 1994, length 18 mm. All specimens were collected in the Chowan River Formation, at the PotashCorps Mine (old Lee Creek Mine), Aurora, Beaufort County, North Carolina, late Piacenzian Pliocene. These species were also present in the Waccamaw Subsea of the Charleston Sea.

Figure 7.14 Gastropods of the late Piacenzian Croatan Subsea, Albemarle Sea.
A= *Brachysycon canaliferum* (Conrad, 1862), length 136 mm; **B**= *Ptychosalpinx tuomeyi* (H. Lea, 1843), length 72 mm; **C, D**= *Volutifusus typus* Conrad, 1863, length 142 mm; **E**= *Ventrilia betsiae* (Olsson and Petit, 1964), length 38 mm; **F, G**= *Urosalpinx suffolkensis* Gardner, 1948, length 32 mm; **H**= *Trossulasalpinx moniliferus* (Emmons, 1858), length 18 mm. All specimens were collected in the Chowan River Formation, at the PotashCorps Mine (old Lee Creek Mine), Aurora, Beaufort County, North Carolina, late Piacenzian Pliocene. These species were also present in the Waccamaw Subsea of the Charleston Sea.

CHAPTER 8. Pleistocene Molluscan Faunas of the Eastern United States

Defined by the extinction of 30% of all marine mollusks worldwide, the Pleistocene was determined to have started approximately 2.58 Ma ago. The 30% extinction level, defined by statistical methods using Lyellian Percentages (Stanley *et al.* 1980), is especially obvious at the stratotype fossil beds on Sicily, Italy. This prominent and easily recognizable faunal impoverishment reflects a cold climatic episode and sea level drop at the very end of the Piacenzian Age of the Pliocene. Along the eastern coast of the United States, the Pliocene-Pleistocene Boundary is marked by the disappearance of most of the index species from the Caloosahatchian Province (listed and discussed in the previous chapter). Immediately following this time of cold climates and lowered sea level, a warm climatic interval occurred, and this new warm-temperate marine world allowed many of the survivors from the Caloosahatchian Province to persist and flourish into the Gelasian Age of the Early Pleistocene. At this time, the enclosed and sheltered Okeechobean Sea of the Everglades Region housed tropical ecosystems that contained classic warm-water gastropod groups such as the cowries (Cypraeidae; see Petuch, Berschauer, and Myers, 2018), vase and chank shells (Turbinellidae), crown conchs (Melongenidae), and giant miters (Pleioptygmatidae), and bivalve groups such as the spiny oysters (Spondylidae). These high tropical (eutropical) mollusks are missing from fossil beds north of southern Florida. The Gelasian molluscan assemblages of the Carolinas and Chesapeake Bay are composed entirely of temperate-water gastropods and are dominated by large gastropods in the family Busyconidae and large pectinid bivalves in the genus *Carolinapecten*.

By the end of the Gelasian Age, the **Waccamaw Subsea** of the Charleston Sea was greatly diminished in size and depth and graded directly into the **Nashua Lagoon System** of the eastern Florida and Georgia coasts (Figure 8.1). At this time, the Albemarle Sea had virtually disappeared and was present only as a shallow indentation of the coastline. The Nashua Lagoon System and the coastal lagoons of the Waccamaw Subsea contained the typical nearshore environments of the Gelasian East Coast of the United States. Extending northward to at least New York, these intertidal areas had muddy sand seafloors and were lined with *Spartina* grass meadows, similar to what is found in the modern Virginian Province. Large busyconid whelks in the genera *Busycon*, *Sinistrofulgur*, *Brachysycon*, and *Busycotypus* occurred in immense aggregations on these mud flats, where they fed upon the extensive beds of shallow infaunal bivalves such as those formed by the venerid *Mercenaria*. These coastal lagoons acted as a refugium for the very last Caloosahatchian Province endemics, including the gastropod such as the muricids *Pterorhytis* and *Mariasalpinx*, the busyconids *Pyruella* and *Brachysycon*, and the cone shell *Ximeniconus*, and bivalves such as the pectinid *Carolinapecten*. These all became extinct along the eastern United States during the late Gelasian Age and their demise heralded the destruction of an entire marine molluscan province. The abrupt removal of so many older Pliocene survivors from the local ecosystems was referred to as the **Caloosahatchian Extinction** by the senior author (Petuch, 2004:260). The following is a partial list of some of the most important and prominent molluscan genera that disappeared from the fossil record during the Caloosahatchian Extinction:

DOI: 10.1201/9781003732983-8

Figure 8.1 Map of the Nashua Lagoon System and eastern coast of the United States during the Gelasian Age of the Early Pleistocene.

EAST COAST GENERA THAT BECAME EXTINCT DURING THE LATE GELASIAN

Gastropoda
Muricidae
Mariasalpinx Petuch, 1988
Pterorhytis Conrad, 1863 (regionally extinct; one living species found off Ecuador)
Trossulasalpinx Petuch, 1986
Fasciolariidae
Pliculofusus Snyder, Vermeij, and Lyons, 2012
Busyconidae
Brachysycon Petuch, 1994
Pyruella Petuch, 1982
Conidae
Contraconus Olsson and Harbison, 1953
Ximeniconus Emerson and Old, 1962 (regionally extinct; still living in the Eastern Pacific)
Bivalvia
Glycymeridae
Costaglycymeris Ward, 1992
Ostreidae
Conradostrea Ward and Blackwelder, 1987
Pectinidae
Carolinapecten Ward and Blackwelder, 1987
Stralopecten Tucker-Rowland, 1938
Crassatellidae
Marvacrassatella Ward and Blackwelder, 1987

Fauna of the Late Pleistocene Carolinian Province

Following the cold climate event and lowered sea level during the end of the Gelasian Age, warmer water conditions returned to the eastern United States during the Calabrian Age of the Middle Pleistocene (1.8 Ma). The physiologically hardy survivors of the Caloosahatchian Extinction went on to establish entire new ecosystems all along the East Coast and laid the foundations for a new bioregion, the **Carolinian Molluscan Province** (described in Chapters 1 and 2 of this book). As conditions warmed from Cape Hatteras southward, classic Carolinian index taxa, such as the large muricid *Hexaplex fulvescens* (Figure 8.7A, B) migrated into the new provincial boundaries and quickly became established as a dominant drilling predator on other mollusks. Based on data from the Socastee Formation and other late Pleistocene fossil beds in the Carolinas, a full complement of Carolinian Province mollusks had become established by the Holocene Boundary, including classic Carolinian species such as the fasciolariids *Cinctura hunteria* and *Triplofusus giganteus*, and the olivid *Americoliva sayana* (all described and illustrated in Richards, 1962). Although the molluscan assemblages of late Pleistocene resemble impoverished versions of the modern East Coast fauna, they did contain one problematical species of great interest, Stone's Neptune shell, *Neptunea (Atractodon) stonei* (Pilsbry, 1893) (shown here on Figure 8.7 C, D). This unusual buccinid has been found in fossil beds from Nova Scotia, Canada to Cape Hatteras, North Carolina and is the only species of late Pleistocene gastropod that is known to have become extinct at the beginning of the Holocene.

The Champlain Sea

Named for Lake Champlain of New York and Vermont States and the Canadian Province of Quebec, the **Champlain Sea** was the northernmost of the eastern North American paleoseas (Figure 8.2). This long, narrow body of sea water was a giant inlet off the North Atlantic Ocean that flooded inland due to the continental landmass being compressed downward into the Earth's mantle. The extreme isostatic down-warping of the entire St. Lawrence River Valley and Lake Champlain area was due to the immense weight of the Laurentide Glacial Ice Sheet just to the north. At its peak development, the Champlain Sea extended as far west as modern Lake Ontario and the Ottawa River, as far east as the Gulf of St. Lawrence, and encompassed all of Lake Champlain. The fossil beds found along the St. Lawrence River, eastern Lake Ontario, and around Lake Champlain (the Riviere-du-Loup and Champlain Formations and the Prescott Beds) have revealed that the Champlain Sea contained an impoverished high arctic-boreal molluscan fauna (some examples shown here on Figures 8.8, 8.9, and 8.10; see Richards, 1962). As an arctic sea bordered by glacial ice shelves, the Champlain Sea was also a home to large pods of Beluga Whales (*Delphinapterus leucas*), and their subfossil bones can be collected in the fossil beds around modern Lake Champlain and the Thousand Islands of eastern Lake Ontario.

The molluscan fauna of the Champlain Sea, although impoverished, contained several interesting species and variants. Of special interest is the astartid bivalve, *Astarte laurentiana* Lyell, 1845 (Figure 8.8F), which is the only known species of mollusk that was endemic to the Champlain Sea (**authors' note**: this interesting little endemic clam was discovered and named by Sir Charles Lyell, the "Father of Modern Geology", during his fourth research trip to North America). Other noteworthy shells from the Champlain Sea include two species of dwarf neptune shells, averaging only around 55 mm in length. These tentatively have been identified as *Neptunea despecta* and *Neptunea vindlandica* but exhibit several morphological differences and may, with more study, prove to be new endemic Champlain subspecies. For details on the ecology of the Champlain molluscan communities, see Petuch, 2004:253-258. During the warming at the Pleistocene-Holocene Boundary around 11,700 years BP, the great Laurentide Ice Sheets began to melt, and the continental glaciers started to retreat. This abrupt loss of weight on the continental plate caused isostatic glacial rebound to take place and the old seafloors rose to heights above sea level. At this time, the Champlain Sea drained off the North American continent, and the marine conditions were replaced by freshwater rivers and lakes. The basin of the Champlain Sea gave rise to modern eastern Lake Ontario, Lake St. Louis, the St. Lawrence River and Seaway, Lake Champlain, Lake George, and the Gulf of St. Lawrence. Rock nodules and concretions collected along Frontenac Island in eastern Lake Ontario often contain fossil Three-Spined Stickleback Fishes (*Gasterosteus aculeatus*), demonstrating that the western Champlain Sea supported large estuarine environments and brackish water lagoons.

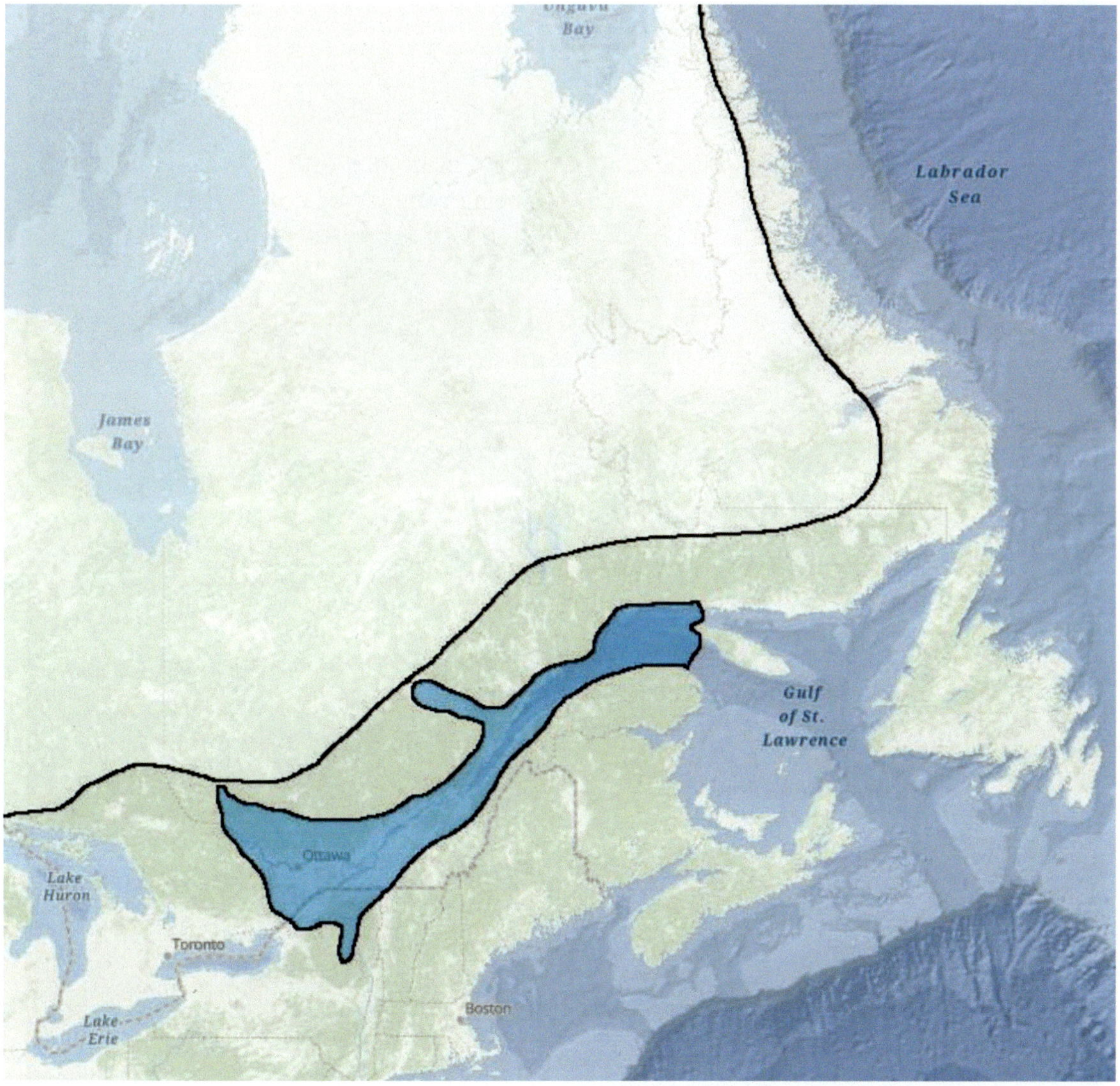

Figure 8.2 Map of the Champlain Sea showing its proximity to the Laurentide Ice Sheet. Blue color denotes the extent of the Champlain Sea; the light colored area to the north denotes the eastern edge of the Laurentide Ice Sheet.

Iconography of Eastern United States Pleistocene Mollusks

Figure 8.3 Mollusks of the Gelasian Coastal Lagoon Systems.
A, B= *Volutifusus halscotti* Petuch, 1994, length 115 mm; **C=** *Triplofusus acmaensis* (B. Smith, 1940), length 358 mm, with growths of the coral *Septastrea*; **D=** *Busycon rucksorum* Petuch, 1994, length 215 mm; **E=** *Sinistrofulgur yeehaw* Petuch, 1994, length 152 mm; **F=** *Hexaplex trippae* Petuch, 1994, length 110 mm. Collected in coastal lagoon deposits of the Waccamaw and Nashua Formations, late Gelasian Pleistocene of Florida and the Carolinas.

Figure 8.4 Gastropods of the Gelasian Waccamaw Subsea, Charleston Sea.
A= *Heilprinia equalis* (Emmons, 1858), length 76 mm; **B, C=** *Urosalpinx rucksorum* Petuch, 1994, length 21 mm; **D=** *Americoliva rucksorum* (Petuch, 1994), length 38 mm; **E=** *Pyruella bladenense* (Gardner, 1948), length 112 mm; **F=** *Ximeniconus waccamawensis* (B. Smith, 1930), length 35 mm; **G=** *Carolinapecten senescens* (Dall, 1898), length 87 mm. All specimens were collected in the Waccamaw Formation, late Gelasian Pleistocene, at the Calabash Pit, Shallotte, Brunswick County, North Carolina. Also found in the contemporaneous Nashua Formation of coastal eastern Florida.

Figure 8.5 Gastropods of the Gelasian Waccamaw Subsea, Charleston Sea.
A= *Cinctura rucksorum* (Petuch, 1994), length 115 mm; **B=** *Scaphella brennmortoni* Olsson and Petit, 1964, length 98 mm; **C, D=** *Mariasalpinx auroraensis* Petuch, 1994, length 36 mm; **E=** *Sinistrofulgur pamlico* Petuch, 1994, length 96 mm; **F=** *Americoliva roseae* (Petuch, 1991), length 84 mm; **G=** *Nodipecten vaccamavensis* (Olsson, 1914), length 74 mm. All specimens were collected in the Waccamaw Formation, late Gelasian Pleistocene, at the Calabash Pit, Shallotte, Brunswick County, North Carolina. Also found in the contemporaneous Nashua Formation of coastal eastern Florida and the James City Formation of eastern North Carolina and Virginia.

Figure 8.6 Mollusks of the Gelasian Waccamaw Subsea, Charleston Sea.
A= *Brachysycon willcoxi* (Gardner, 1948), length 114 mm; **B**= *Busycotypus scotti* Petuch, 1994, length 169 mm; **C**= *Busycon gilmorei* Petuch, 1994, length 165 mm; **D**= *Brachysycon amoenum* (Conrad, 1875), length 115 mm; **E**= *Carolinapecten planicostata* (Gardner, 1943), length 175 mm. All specimens were collected in the Waccamaw Formation, late Gelasian Pleistocene, at the Calabash Pit, Shallotte, Brunswick County, North Carolina. Also found in the contemporaneous Nashua Formation of coastal eastern Florida and the James City Formation of eastern North Carolina and Virginia.

Figure 8.7 Gastropods from Late Pleistocene-Holocene Beds.
A, B= *Hexaplex fulvescens* (Sowerby II, 1834), length 143 mm, found in latest Pleistocene-Holocene fossil beds all along the eastern United States coast, from North Carolina to Florida; **C, D=** *Neptunea (Atractodon) stonei* (Pilsbry, 1893), syntype specimen, length 82 mm, found in the late Pleistocene Cape May Formation (second terrace) at Cape May, Cape May County, New Jersey. This unusual buccinid has been found in fossil beds from Nova Scotia, Canada to Cape Hatteras, North Carolina. This is the only species of late Pleistocene gastropod known to have become extinct at the beginning of the Holocene.

Figure 8.8 Mollusks of the Late Pleistocene-Holocene Champlain Sea.
A, B= *Buccinum glaciale donovani* Gray, 1824, length 41 mm; **C, D=** *Buccinum fragile* Verkreuzen, 1878, length 44 mm; **E=** *Cryptonatica clausa* (Broderip and Sowerby, 1829), length 26 mm; **F=** *Astarte laurentiana* Lyell, 1824, length 20 mm; **G, H=** *Boreoscala greenlandicum* (Perry, 1811), length 38 mm. All specimens were collected in a sand quarry at St. Nicholas, Quebec, Canada, Riviere-du-Loup Formation, early Holocene. These species are typical of the Champlain glacial sea fauna that inhabited present-day Lake Champlain and the St. Lawrence River area. The extinct *Astarte laurentiana* is the only mollusk known to have been endemic to the Champlain Paleosea.

Figure 8.9 Neptunes of the Late Pleistocene-Holocene Champlain Sea.
A, B= *Neptunea despecta* (Linnaeus, 1758), length 53 mm; **C, D=** *Neptunea vinlandica* Fraussen and Terryn, 2007, length 56 mm. Both specimens were collected in a sand quarry at St. Nicholas, Quebec, Canada, Riviere-du-Loup Formation, early Holocene. These neptunes were dominant predators on the large bivalve fauna that inhabited the area of present-day Lake Champlain and the St. Lawrence River. These dwarf Neptunes may prove to be undescribed endemic Champlain Sea subspecies.

Figure 8.10 Mollusks of the Late Pleistocene-Holocene Champlain Sea.
A= *Astarte banksii* (Leach, 1819), length 19 mm; **B=** *Hiatella arctica* (Linnaeus, 1767), length 24 mm; **C, D=** *Acirsa borealis* (Lyell, 1842), length 28 mm; **E=** *Trichotropis borealis costellata* Couthouy, 1838, length 10 mm; **F=** *Mya arenaria* Linnaeus, 1758, length 41 mm; **G=** *Macoma balthica* (Linnaeus, 1758), length 16 mm; **H=** *Mya truncata* Linnaeus, 1758, length 42 mm. All specimens were collected in a sand quarry at St. Nicholas, Quebec, Canada, Riviere-du-Loup Formation, early Holocene. These species are typical of the Champlain Sea boreal fauna that inhabited present-day Lake Champlain and the St. Lawrence River area.

View of an exposure of the Nashua Formation in a quarry near Fort Pierce, Florida, showing a molluscan assemblage that is typical of the Gelasian Pleistocene Nashua Lagoon System. Here several large busyconid whelks dominate the paleo-ecosystem, including *Busycon rucksorum*, *Busycotypus scotti*, *Brachysycon amoenum*, and *Sinistrofulgur yeehaw*.

Literature Cited

Abbott, D.H., A.A. Nunez, I.S. Leung, and J.T. Hagstrom. 2003. The Ewing Impact Structure progress report. Abstract #P52A-0472. Fall Meeting 2003. American Geophysical Union.

Bieler, R. and R.E. Petit. 2005. Catalog of Recent and fossil taxa of the family Architectonicidae Gray, 1850 (Mollusca: Gastropoda). *Zootaxa* 1101 (1): 1-119 (December, 2005).

Bousfield, E.L. 1960. *Canadian Atlantic Sea Shells*. National Museum of Canada, Ottawa, Canada. 72 pp.

Campbell, L.D. 1993. *Pliocene Molluscs from the Yorktown and Chowan River Formations of Virginia*. Virginia Division of Mineral Resources, Publication 127. Charlottesville, Virginia. 259 pages.

Collins, L.S., A.G. Coates, W.A. Berggren, M-P. Aubrey, and J. Zhang. 1996. The late Miocene Panama isthmian strait. *Geology* 24: 687-699.

Duncan, C.S. and J. Goff. 2001. Relict iceberg keel marks on the New Jersey outer shelf, southern Hudson Apron. *Geology* 29 (5): 411-414.

Emerson, W.K and M.K. Jacobson. 1971. *Shells from Cape Cod to Cape May, with special reference to the New York City area*. Dover Publications, New York. 152 pp.

Emerson, W.K and M.K. Jacobson. 1976. *The American Museum of Natural History Guide to Shells: Land, Freshwater, and Marine, from Nova Scotia to Florida*. Alfred A. Knopf, New York. 300 pp.

Farley, K.A., D. Vokrouhlicky, D. Bottke, and W, Nesvorny. 2006. A late Miocene dust shower from the break-up of an asteroid in the main belt. *Nature* 439: 295-297.

Glickson, A. 2005. Asteroid/Comet impact clusters, flood basalts, and mass extinctions: significance of isotopic age overlaps. *Earth and Planetary Science Letters* 236: 933-937.

Gorter, J.D. and S.W. Bayford. 2000. Possible impact origin of the middle Miocene (Serravallian) Puffin Structure, Ashmore Platform, northwest Australia. *Australian Journal of Earth Science* 47: 707-714.

Gurov, E.P. and H. P. Gurova. 1993. The Kara-Kul Depression in the Pamirs—a reliable impact structure. *Geologia Zhurnal* 6: 53-64.

Gurov, E.P., H.P. Gurova, R.B. Rakitskaya, A.Y. Yamnichenko. 1993. The Karakul Depression in the Pamirs—the first impact structure in Central Asia (Abstract). *Lunar and Planetary Science* XXIV: 591-592.

Jacobson, M.K. and W.K. Emerson. 1961. *Shells of the New York City Area*. Argonaut Books, Inc., Larchmont, New Jersey. 142 pp.

Leung, I.S, R. Winston, and D.H. Abbott. 2004. Silicon carbide from the Canyon Diablo meteorite and the Ewing Impact Structure. Abstract #P33C-10, Spring Meeting 2004, American Geophysical Union.

Macintyre, I.G. and O.H. Pilkey. 1969. Tropical Reef Corals: Tolerance of Low Temperatures on the North Carolina Continental Shelf. *Science* (17 October 1969), volume 166 issue 3903: 374-375.

Mosher, K. 2024. *Seashells of North Carolina: Revised and Expanded Edition*. Sea Grant North Carolina, The University of North Carolina Press, Chapel Hill, North Carolina. 237 pp.

Oleinik, A.E., L. Marinkovich, K.B. Barinov, and P.K. Swart. 2008. Magnitude of middle Miocene warming in north Pacific high latitudes: stable isotope evidence from *Kaneharaia* (Bivalvia: Dosinidae). *Bulletin of the Geologic Survey of Japan* 59: 339-353.

Petuch, E.J. 1972. *Strombus costatus* and *Morum dennisoni* collected off the North Carolina Coast. *The Veliger* 15: 51-53.

Petuch, E.J. 1974a. A new *Terebra* from the coral reef areas off North Carolina. *The Veliger* 17 (2); 205-208.

Petuch, E.J. 1974b. Two new Pacific Cone Shells (Gastropoda Conidae) and a new *Pleurotomella* from the Hatteras Abyssal Plain. *The Veliger* 17 (3): 40-43.

Petuch, E.J. 1982. Geographical Heterochrony: contemporaneous co-existence of Neogene and Recent molluscan faunas in the Americas. *Palaeogeography, Palaeoclimatology, and Palaeoecology* 37: 277-312.

Petuch, E.J. 1988. *Neogene History of Tropical American Mollusks: biogeography and evolutionary patterns of tropical Western Atlantic Mollusca.* The Coastal Education and Research Foundation, Charlottesville, Virginia. 217 pp.

Petuch, E.J. 1993. Patterns of diversity and extinction in Transmarian Muricacean, Buccinacean, and Conacean Gatropods. *The Nautilus* 106 (4): 155-173.

Petuch, E.J. 1995. Molluscan Diversity in the late Neogene of Florida: evidence for a two-staged mass extinction. *Science* 270: 275-277.

Petuch, E.J. 2004. *Cenozoic Seas: The View from Eastern North America.* CRC Press, Boca Raton, London, New York, 308 pp.

Petuch, E.J. 2012. Miocene Asteroid Impacts: Proposed Effects on the Biogeography and Extinction Patterns of Eastern North American Gastropods. In: *Earth and Life: Global Biodiversity, Extinction Intervals, and Biogeographic Perturbations Through Time.* J.A. Talent, Editor. Special Publication for the International Year of Planet Earth, Springer Publishers, Dordrecht, the Netherlands. 967-981

Petuch, E.J. 2013. *Biogeography and Biodiversity of Western Atlantic Mollusks.* CRC Press, Boca Raton, London, New York. 234 pp.

Petuch, E.J. and D.P. Berschauer. 2021. *Tropical Marine Mollusks: An Illustrated Biogeographical Guide.* CRC Press, Boca Raton, London, New York. 357 pp.

Petuch, E.J. and D.P. Berschauer. 2022. *Ancient Seas of Southern Florida: The Geology and Paleontology of the Everglades Region.* CRC Press, Boca Raton, London, New York. 266 pp.

Petuch, E.J. and D.P. Berschauer. 2023. *The Ecphoras: Iconic Fossils of Eastern North America.* CRC Press, Boca Raton, London, New York. 207 pp.

Petuch, E.J. and D.P. Berschauer. 2025. *Molluscan Faunas of the Gulf of Mexico: Endemism in North America's Inland Sea.* CRC Press, Boca Raton, London, New York. 223 pp.

Petuch, E.J. and M. Drolshagen. 2009. *Molluscan Paleontology of the Chesapeake Miocene.* CRC Press, Boca Raton, London, New York. 152 pp.

Petuch, E.J. and R, F. Myers. 2014. *Molluscan Communities of the Florida Keys and Adjacent Areas: Their Ecology and Biodiversity.* CRC Press, Boca Raton, London, New York. 300 pp.

Petuch, E.J., Myers, R.F, and D.P. Berschauer. 2015. *The Living and Fossil Busycon Whelks: Iconic Mollusks of Eastern North America.* Special Publication of the San Diego Shell Club, Encinitas, California. 194 pp.

Petuch, E.J. and C. Roberts. 2007. *The Geology of the Everglades and Adjacent Areas.* CRC Press, Boca Raton, London, New York. 212 pp.

Petuch, E.J. and D.M Sargent.2011. *Rare and Unusual Shells of the Florida Keys and Adjacent Areas.* MdM Publishing, Wellington, Florida. 159 pp.

Richards, H.G. 1962. Studies on the Marine Pleistocene: Part 1, The Marine Pleistocene of the Americas and Europe; Part 2, The Marine Pleistocene Mollusks of Eastern North America. *Transactions of the American Philosophical Society* New Series 52 (3): 1-141.

Shoemaker, E.M. and E.C.T. Chao. 1961. New Evidence for the impact origin of the Ries Basin, Bavaria, Germany. *Journal of Geophysical Research* 66: 3371-3378.

Stanley, S.M., W.O. Addicott, and K. Chinzei. 1980. Lyellian curves in paleontology: Possibilities and limitations. *Geology* (1980) 8 (9): 422-426.

SYSTEMATIC INDEX

Only the gastropods and bivalves that are illustrated in the iconographies at the end of each chapter are listed here. These are arranged here by genus, both living and fossil species.

A

B

C

K

L

M

N

O

P

R

S

U

V

X

OCEANOGRAPHIC AND GEOLOGICAL INDEX

P

R

S

T

V

W

Y

View from Pinckney Island, South Carolina. These marshes are typical of the coastal environments of the southeastern United States.

About the Authors

Edward J. Petuch was born in Bethesda, Maryland, in 1949. Raised in a Navy family, he spent many of his childhood years collecting living and fossil shells in such varied localities as Chesapeake Bay, California, Puerto Rico, and Wisconsin. His early interests in malacology and oceanography eventually led to BA and MS degrees in Zoology from the University of Wisconsin-Milwaukee. During his Master's thesis research, Petuch concentrated on the molluscan biogeography of West Africa, traveling extensively in the Canary Islands, Western Sahara, Senegal, Gambia, Sierra Leone, Ivory Coast, and Cameroon. During this time, he also conducted research on the molluscan ecology of both coasts of Mexico and the Great Barrier Reef of Belize.

Continuing his education, Petuch studied marine biology and malacology under Gilbert Voss and Donald Moore at the Rosenstiel School of Marine and Atmospheric Sciences at the University of Miami, where he received a full scholarship. During this time, his doctoral dissertation research involved intensive field work in Costa Rica, Colombia, Venezuela, Barbados, the Grenadines, and Brazil, where he often went to sea with the local shrimpers for weeks at a time. After receiving his PhD in oceanography in 1980, Petuch was invited to conduct two years of postdoctoral research, funded by the National Science Foundation, with Geerat Vermeij at the University of Maryland. While there, he also held a research associateship with the Department of Paleobiology at the National Museum of Natural History, Smithsonian Institution, under the sponsorship of Thomas Waller, and conducted field work in the Pliocene and Pleistocene fossil beds of Florida and North Carolina and the Miocene fossil beds of Maryland and Virginia.

Petuch has also collected and studied living mollusks in Australia, Papua-New Guinea, Fiji, French Polynesia, Japan, the Bahamas, Nicaragua, and Uruguay. This research has led to the publication of over 500 scientific papers and the discovery and description of over 2,000 new species of mollusks and 200 new genera. His previous 27 books are well-known reference texts in the malacological and paleontological communities, and some of the more recent better-known ones include: *Molluscan Faunas of the Gulf of Mexico: Endemism in North America's Inland Sea* (2025); *The Ecphoras: Iconic Fossils of Eastern North America* (2023); *Mollusks and Marine Environments of the Ten Thousand Islands* (2022); *Ancient Seas of Southern Florida: The Geology and Paleontology of the Everglades Region* (2022); *Tropical Marine Mollusks: An Illustrated Biogeographical Guide* (2021); *Jewels of the Everglades: The Fossil Cowries of Southern Florida* (2018); *The Living and Fossil Busycon Whelks: Iconic Mollusks of Eastern North America* (2015); *Cone Shells of the Okeechobean Sea* (2015); *Molluscan Communities of the Florida Keys and Adjacent Areas: Their Ecology and Biodiversity* (2014); *Biogeography and Biodiversity of Western Atlantic Mollusks* (2013); *Molluscan Paleontology of the Chesapeake Miocene* (2010); *The Geology of the Everglades and Adjacent Areas* (2007); and *Cenozoic Seas: The View from Eastern North America* (2004).

Currently, Petuch is a Professor Emeritus in the Department of Geosciences, Florida Atlantic University in Boca Raton, Florida where, for thirty years, he taught undergraduate classes in oceanography, paleontology, and physical geology, and graduate classes in paleoecology and paleoceanography. He currently resides in Jupiter, Florida, with his wife Linda, where they both enjoy visits from their three children and their families.

© Morgan Taylor Photography, with permission

David P. Berschauer was born in Rockville Center, New York, in 1964. He spent his youth collecting shells in such varied localities as New York, Florida, California, Washington, Mexico, and throughout the Caribbean. His early interests in natural history, malacology, and marine biology eventually led to a BS in biology at the University of California-Irvine, an advanced marine invertebrate zoology course at Washington State University's Friday Harbor Marine Lab, and studies toward the pursuit of a graduate degree in marine biology at Florida State University in Tallahassee, Florida. While still a college undergraduate, Berschauer performed field biology research, published a number of research papers, and gave scientific presentations at national conferences.

He subsequently switched career paths and attended Southwestern University School of Law in Los Angeles, California, earning his Juris Doctorate in 1991. Although having developed a legal career as a trial lawyer spanning over thirty-five years and is licensed in both California and Georgia, he has kept malacology as a lifetime avocation and has put together a sizeable research collection and personal museum of molluscan specimens.

Over his entire professional life, Berschauer continued to pursue his passion for marine biology, and collecting and studying marine organisms. In his spare time, he has developed and published a relational database software program to aid in the organization and maintenance of a systematic collection. Although originally designed for malacology, the program is applicable to entomology and other aspects of systematic zoology.

Berschauer is an active member of the San Diego Shell Club, a member of the Gulf Coast Shell Club, a Museum Associate with the Natural History Museum of Los Angeles County (LACM) Malacology Department, a Collection Associate with the Santa Barbara Museum of Natural History (SBMNH), and is the Editor-in-Chief of the journal, *The Festivus*. He is also well-known for his natural history and shell photography, with a multitude of high-quality examples seen throughout this book. Besides being the author of over 90 important scientific papers on molluscan systematics and ecology, Berschauer has described and named over 150 new species of gastropods and is also the co-author of several recent books on mollusks, including: *The Living and Fossil Busycon Whelks: Iconic Mollusks of Eastern North America* (2015); *Jewels of the Everglades: The Fossil Cowries of Southern Florida* (2018); the well-received *Sea Shells of Southern California: Marine Shells of the Californian Province* (2018); *Tropical Marine Mollusks: An Illustrated Biogeographical Guide* (2021); *Ancient Seas of Southern Florida: The Geology and Paleontology of the Everglades Region* (2022); *Mollusks and Marine Environments of the Ten Thousand Islands* (2022); *The Ecphoras: Iconic Fossils of Eastern North America* (2023); and *Molluscan Faunas of the Gulf of Mexico: Endemism in North America's Inland Sea* (2025). He currently resides in Cairo, Georgia, with his wife, Felicia, where they both enjoy visits with their children and their spouses, and their four young grandchildren.